MANUFACTURING ENGINEERING EDUCATION

MANUFACTURING ENGINEERING EDUCATION

Edited by

J PAULO DAVIM

CHANDOS
PUBLISHING

An imprint of Elsevier

Chandos Publishing is an imprint of Elsevier
50 Hampshire Street, 5th Floor, Cambridge, MA 02139, United States
The Boulevard, Langford Lane, Kidlington, OX5 1GB, United Kingdom

Notices
Knowledge and best practice in this field are constantly changing. As new research and experience
broaden our understanding, changes in research methods, professional practices, or medical treatment
may become necessary.

Practitioners and researchers must always rely on their own experience and knowledge in evaluating
and using any information, methods, compounds, or experiments described herein. In using such
information or methods they should be mindful of their own safety and the safety of others,
including parties for whom they have a professional responsibility.

To the fullest extent of the law, neither the Publisher nor the authors, contributors, or editors,
assume any liability for any injury and/or damage to persons or property as a matter of products liability,
negligence or otherwise, or from any use or operation of any methods, products, instructions, or
ideas contained in the material herein.

Library of Congress Cataloging-in-Publication Data
A catalog record for this book is available from the Library of Congress

British Library Cataloguing-in-Publication Data
A catalogue record for this book is available from the British Library

ISBN: 978-0-08-101247-5
ISBN: 978-0-08-101264-2

For information on all Chandos publications visit our
website at https://www.elsevier.com/books-and-journals

Working together
to grow libraries in
developing countries

www.elsevier.com • www.bookaid.org

Publisher: Glyn Jones
Acquisition Editor: Glyn Jones
Editorial Project Manager: Joshua Bayliss
Production Project Manager: Sojan P. Pazhayattil
Cover Designer: Vistoria Pearson

Typeset by SPi Global, India

CONTENTS

CONTRIBUTORS

A.K. Basak
Adelaide Microscopy, The University of Adelaide, Adelaide, SA, Australia

N. Bay
Department of Mechanical Engineering, Technical University of Denmark, Lyngby, Denmark

P. Christiansen
Department of Mechanical Engineering, Technical University of Denmark, Lyngby, Denmark

Andrea Corrado
Department of Civil and Mechanical Engineering, Università di Cassino e del Lazio Meridionale, Cassino, Italy

Suman Kalyan Das
Department of Mechanical Engineering, Jadavpur University, Kolkata, India

Uday S. Dixit
Department of Mechanical Engineering, Indian Institute of Technology Guwahati, Guwahati, India

Y. Dong
School of Civil and Mechanical Engineering, Curtin University, Bentley, WA, Australia

Manjuri Hazarika
Department of Mechanical Engineering, Assam Engineering College, Guwahati, India

M.N. Islam
School of Civil and Mechanical Engineering, Curtin University, Bentley, WA, Australia

Nikolaos Kontogiannis
National Technical University of Athens, School of Mechanical Engineering, Section of Manufacturing Technology, Athens, Greece

P.A.F. Martins
IDMEC, Instituto Superior Tecnico, University of Lisbon, Lisbon, Portugal

J. Paulo Davim
Department of Mechanical Engineering, University of Aveiro, Aveiro, Portugal

Wilma Polini
Department of Civil and Mechanical Engineering, Università di Cassino e del Lazio Meridionale, Cassino, Italy

A. Pramanik
School of Civil and Mechanical Engineering, Curtin University, Bentley, WA, Australia

Prasanta Sahoo
Department of Mechanical Engineering, Jadavpur University, Kolkata, India

George-Christopher Vosniakos
National Technical University of Athens, School of Mechanical Engineering, Section of Manufacturing Technology, Athens, Greece

Nikolaos Zourtsanos
National Technical University of Athens, School of Mechanical Engineering, Section of Manufacturing Technology, Athens, Greece

ABOUT THE EDITOR

J. Paulo Davim received a Ph.D in mechanical engineering in 1997, a M.Sc. in mechanical engineering (materials and manufacturing processes) in 1991, his Mechanical Engineering degree (5 years) in 1986, from the University of Porto (FEUP), his Aggregate title (Full Habilitation) from the University of Coimbra in 2005, and his D.Sc. from London Metropolitan University in 2013. He is Eur Ing by FEANI-Brussels and senior chartered engineer by the Portuguese Institution of Engineers with a MBA and specialist title in Engineering and Industrial Management. Currently, he is a professor at the Department of Mechanical Engineering of the University of Aveiro, Portugal. He has more than 30 years of teaching and research experience in Manufacturing, Materials and Mechanical Engineering with special emphasis in Machining & Tribology. He has also interest in Management & Industrial Engineering and Higher Education for Sustainability & Engineering Education. He has guided large numbers of postdoc, Ph.D, and masters students as well as coordinated and participated in several financed research projects. He has received several scientific awards. He has worked as evaluator of projects for international research agencies as well as examiner of Ph.D thesis for many Universities. He is the editor-in-chief of several international journals, guest editor of journals, books editor, book series editor, and scientific advisory for many international journals and conferences. Presently, he is an editorial board member of 25 international journals and acts as a reviewer for more than 80 prestigious Web of Science journals. In addition, he has also published as editor (and coeditor) more than 100 books and authored (and coauthored) more than 10 books, 70 book chapters, and 400 articles in journals and conferences (more than 200 articles in journals indexed in Web of Science core collection/ h-index 47 +/6500 + citations and SCOPUS/h-index 53 +/9000 + citations).

J. Paulo Davim
Aveiro, Portugal

PREFACE

Nowadays, *Manufacturing Engineering* is defined as a discipline *"which involves the ability to plan the processes and practices of manufacturing and to research and develop systems, processes, machines, tools and equipment for producing quality products"*. Recently, manufacturing engineering also emphasized on some modern subjects such as nanomanufacturing, biomanufacturing as well as aspects related to sustainable manufacturing. Manufacturing engineering is related to mechanical, industrial, and production engineering.

The book aims to provide information on manufacturing engineering education for modern industry. The initial chapter of the book provides history of production and industrial engineering through contributions of stalwarts. Chapter 2 is dedicated to manufacturing engineering education (Indian perspective). Chapter 3 presents learning enhancement of project-based unit in mechanical engineering undergraduate course. Chapter 4 covers friction compensation in the compression test. Chapter 5 contains appreciation of CNC technology through machine tool upgrading by an open controller. Finally, the last chapter of the book is dedicated to model the assembly of thin parts in composite material.

The present book can be used as a book for final undergraduate engineering course or as a topic on manufacturing engineering at the postgraduate level. This book can serve as a reference for academicians, researchers, manufacturing, mechanical, and industrial engineers as well as professionals in production engineering. Also, this book presents scientific interest for institutes, centers of the research, laboratories, and universities throughout the world.

The editor acknowledges Elsevier for this opportunity and for their professional support. Finally, I would like to thank all the chapter authors for their availability for this work.

J. Paulo Davim
Aveiro, Portugal

CHAPTER 1

History of Production and Industrial Engineering Through Contributions of Stalwarts

Manjuri Hazarika*, Uday S. Dixit†, J. Paulo Davim‡
*Department of Mechanical Engineering, Assam Engineering College, Guwahati, India
†Department of Mechanical Engineering, Indian Institute of Technology Guwahati, Guwahati, India
‡Department of Mechanical Engineering, University of Aveiro, Aveiro, Portugal

1 INTRODUCTION

From time immemorial, human race is in the continuous process of utilizing natural resources for the betterment of the quality of life. Production processes transform resources into useful products. Human beings started manufacturing activities millions of years ago. It is believed that tools were used in Eolithic period that started circa 10 million years before present (McNiel, 1990). Earlier, the human beings used to make the artifacts for their own consumption. They were using muscle power assisted by handheld tools. Gradually, mankind learnt to replace muscle power with animal power and power from nature such as flowing water and wind. Simultaneously, the family system emerged and families formed communities, villages, and towns. With the growth of civilization, the division of work gained importance. However, the manufacturing used to be often an individualistic activity, at most confined to a family. For example, before the onset of the First Industrial Revolution in the 18th century, mainly the domestic system of production was prevalent in England (Bythell, 1983).

Modification of Newcomen's 1712 AD steam engine by James Watt in 1769 AD (Crump, 2007) was a significant event that provided a great impetus to the First Industrial Revolution that lasted till circa 1860 AD. As a result, traditional artisan-based techniques of producing goods were replaced by small workshops and factories. Engineers played a substantial role in starting and sustaining the factory system of production. In the beginning (c.325 AD), the word "engineer" meant constructor of military weapons such as a catapult (Baofu, 2009) and engine was a general term for military weapons. The term Civil Engineering was first coined in the 18th century to

Manufacturing Engineering Education
https://doi.org/10.1016/B978-0-08-101247-5.00001-0

1

indicate engineering related to nonmilitary applications, thus differentiating it from Military Engineering. The first British professional society, Institution of Civil Engineers, was formed (http://www.ice.org.uk, n.d.) in 1818. In 1847, Institution of Mechanical Engineers was formed in the United Kingdom (http://www.imeche.org/, n.d.). Society of Telegraph Engineers was formed in 1871 that was later converted to Institution of Electrical Engineers in 1889. However, it is known as The Institution of Engineering and Technology since 2006 (http://www.theiet.org/resources/library/archives/research/guides-iet.cfm, n.d.). In the United States, American Institute of Electrical Engineers was formed in 1884, which is now known as Institute of Electrical and Electronics Engineers (https://www.ieee.org/index.html, n.d.). Mining and Metallurgical Engineering were born in late 19th century out of Civil and Mechanical Engineering. Chemical Engineering was developed out of applied chemistry and engineering practices.

Second Industrial Revolution started in late 19th century (Mokyr, 1998). More emphasis was paid during this period on optimal utilization of resources and maximizing the profit. The need for increasing efficiency and effectiveness of production methods gave birth to a separate discipline of engineering called Production Engineering. Production Engineers are more concerned with proper and efficient utilization of technology rather than in developing drastically new technologies. Institution of Production Engineers was founded in the United Kingdom in 1921 (Millerson, 1964). Simultaneously in the United States, Industrial Engineering was developed; Hugo Diemer (1870–1937) coined the term "Industrial Engineering" in 1900 in the United States and founded a Department of Industrial Engineering at Penn State (Nanda, 2006). Diemer published the first book on Industrial Engineering entitled "Factory Organization and Administration". Institute of Industrial Engineers was founded in 1948 in the United States, which has been renamed as Institute of Industrial & Systems Engineers (http://www.iise.org/Home/, n.d.).

Nowadays, the term Production Engineering is used interchangeably for Industrial Engineering. In the opinion of the authors of this article, although literally there is a difference between the words "Production" and "Industrial," these branches were developed due to different geographical regions; the term Production Engineering was more prevalent in the United Kingdom and Industrial Engineering in the United States. Globalization mingled both of them. Literally, a Production Engineer is concerned about proper application and utilization of technology for enhancing the production of a factory. However, nowadays, Production Engineers are gaining employment in service sectors such as banking and insurance also because the concepts of production can be applied to service sectors as well. Industrial Engineering, on the other hand, deals with all engineering aspects of industry starting from product design to marketing. It is very close to management and, at several places, the name of the department itself is Industrial Engineering and Management.

This article intends not to distinguish Production and Industrial Engineering in a very formal way. Instead, they are treated as two twin-sisters if not just one entity. Their history is presented through the contributions of the stalwarts. As history is mostly concerned with the affairs related to human beings, the contributions of the leading personalities can throw light on the historical background.

2 PERIOD OF FIRST INDUSTRIAL REVOLUTION (1750–1850)

Before the Industrial Revolution, which began in England during 1750s, goods had been produced in batches and required manual labor in all phases of production. Industrial Revolution and invention of different machines for mechanization of production processes reinforced each other. Mechanization started in England and other parts of Europe with the development of textile machinery (Martin-Vega, 2004). In 1733, Jonh Kay (c.1704 – 1779) developed a flying shuttle that tremendously enhanced the productivity of weaving. James Hargreaves (c.1720–1778) invented a multispindle spinning frame, spinning jenny, in 1765, followed by the water-powered spinning mill by Richard Arkwright (1732–1792) in 1769. Modification of steam engine by James Watt led to the development of a lot of machine tools, for example, a cannon boring mill was developed by John Wilkinson (1728–1808) in 1774, a cylindrical boring machine was developed by John Smeaton (1724–1792), and so on (Dixit et al., 2017). Industrial Revolution soon spread to America, where it continued to be further developed. In the field of Mechanical Engineering, a lot of efforts were made for developing machine tools such as turning, milling, and drilling machines. These machines enhanced the productivity and lessened the importance of manual work. Starting from the later part of the 18th century, specialized engineering workshops expanded their production facilities and started production and installation of steam engines and other machinery (Crump, 2007). A wide range of industries were set up such as cotton and textile industry, paper making industry, and printing industry, which contributed immensely toward the growth of Production and Industrial Engineering. Mechanization of production was the distinct feature of the industry during this period. The following subsections present contributions of some leading personalities of this period and their invaluable contributions.

2.1 Adam Smith (1723–1790)

Seeds of Industrial Engineering were sown by Adam Smith. This economist from Scotland provided strong foundation to the modern economic theory. In 1776, he published his famous book "An Inquiry into the Nature and Causes of the Wealth of Nations" (Capra and Luisi, 2014). The book adopts the theme of laissez faire, which he called "invisible hand" of the market. He believed that the market forces took care of the economy for the betterment of everyone. His book highlights the important concepts of division and specialization of labor for enhancing efficiency and productivity. Smith identified land, labor, and capital as the main inputs to a production system and productivity of labor as the key factor for economic growth of a nation. Motivated by his ideas, many innovators tried to implement the same in factory systems of production during Industrial Revolution. As a result, manual operations were replaced with mechanized operations, thus paving way for the mass production of goods. His concepts of supply and demand were later expanded by economists to develop various economic models.

2.2 Eli Whitney (1765–1825)

The early stages of Production Engineering can be traced back to the period before the First World War. Eli Whitney, an American Mechanical Engineering graduate from Yale, is the pioneer for introducing the idea of mass production of interchangeable parts. Around 1790s, America feared war with Britain and France. The US government signed a contract with Eli Whitney asking him to manufacture about 10,000 muskets for the ensuing war (Armytage, 1961). He invented a technique using jigs for producing interchangeable parts to produce the vast quantities of muskets. Mass production technique of interchangeable parts had revolutionary effect on productivity saving on resources like manpower, time, and cost. He also tried to implement measures for quality control and cost analysis in production. By this time, the idea of factory system of production and mechanization of production processes had spread to America from England as a result of Industrial Revolution. Simeon North (1765 – 1852), another inventor, used the mass production technique to manufacture and supply pistols to the US government for war.

As mentioned earlier, most of the patented production machineries in the early stages of Industrial Revolution were in the textile industry. However, cotton mills in Britain were short of raw materials as manual cleaning of cotton from its seeds was highly time-consuming. Cotton gin, a machine for cleaning cotton, was invented by Eli Whitney in 1793 that replaced manual cleaning of cotton by machine cleaning, which had tremendous effect on the cotton industry. Whitney's cotton gin was made with a hopper to feed the cotton, a rotating cylinder fitted with short wire hooks over which cotton fibers pass, another cylinder with brushes rotating in the opposite direction to brush off the cotton from the wire hooks, and a strainer to take out the seeds (https://www.britannica.com/biography/Eli-Whitney, n.d.). Introduction of water-powered spinning mills by Richard Arkwright gradually replaced manual spinning, thus accelerating the growth of textile industry. A new era started in 1812 when Paul Moody (1779 − 1831) and F.C. Lowell (1775 − 1817) produced cloth by spinning raw cotton in a mill; they were instrumental in bringing Industrial Revolution to America. Thus, mechanization of spinning and weaving took textile industry to a new height. Further developments followed in the ensuing period, resulting in production of goods using the new techniques of production.

2.3 Samual Bentham (1757–1831)

Some of the noteworthy persons who implemented mass manufacturing in England are Samual Bentham, Marc Brunel, and Henry Maudslay. Apart from Eli Whitney, Samual Bentham, the British engineer and naval architect, is also given credit for introducing mass production. Bentham worked in collaboration with Mark Brunel and Henry Maudslay to produce pulley blocks for naval use in "Block mill," a mass production factory conceived by him (https://www.britannica.com/biography/Samuel-Bentham, n.d.). Brunel designed the machineries based on Bentham's idea and Maudslay fabricated them to produce pulley blocks to be used in ships. Bentham's factory was capable of producing around 160,000 interchangeable pulley blocks per year.

2.4 Mark Brunel (1769–1849)

Mark Brunel was a French civil engineer who contributed immensely toward design and construction of ships and railways in their early stages.

Naval Architecture and Engineering reached a new height with implementation of hydrodynamic principles in the design of ships by Mark Brunel. He is considered as the pioneer in building large iron ships considering the size and weight of a ship, its speed, longitudinal strength, and the volume of water displaced. He designed "The Great Eastern", which was the largest steamship until 20th century (Burstall, 1963). Mark Brunel along with Samual Bentham and Henry Maudslay designed a mass production method to produce pulley blocks for ships as mentioned earlier. Mark Brunel is also famous for his contributions to the evolution of railways. In 1841, Brunel designed a locomotive named "The Lord of the Isles" for a British railway company named The Great Western Railway.

2.5 Henry Maudslay (1771–1831)

The contribution of Henry Maudslay toward machine tool technology is indispensable. His inventions were important foundation for the growth of Industrial Revolution. On the onset of Industrial Revolution, Maudslay set up a small workshop with the machine tools he invented and patented. Later, it became one of the most important industries in England by the name "Maudslay, Sons and Field" where some famous inventors, e.g., Richard Roberts (1789 – 1864), Sir Joseph Whitworth (1803 – 1887), and James Nasmyth (1808 – 1890), started their careers. Henry Maudslay is best remembered for his invention of the screw-cutting lathe in 1800, which was capable for cutting a large number of screw threads of the same size. It was the first engine lathe having a guide screw, which started mass production of standardized screw threads. Standard screw threads and nuts allowed the interchangeability during assembly. Mass production of the pulley blocks by Maudslay, Brunel, and Bentham was of great help in the war against Napoleon. The role of Henry Maudslay in the development of machine tool technology is crucial and his inventions were used in the engineering workshops across the world. Later, with the progress of machine tool industry, turret lath was developed for mass production where several tools could be mounted on a turret that could be indexed. Stephen Fitch built the first turret lathe carrying eight tools in c.1845 in the United States (Robert, 1989).

2.6 Charles Babbage (1791–1871)

The British mathematician and engineer, Charles Babbage, was an enthusiastic follower of the principles of Adam Smith on productivity. He visited a number of factories in England and the United States to promote Adam Smith's idea of division and specialization of labor to enhance productivity. He studied the manufacturing operations in the factories, observed, and recorded the minute details such as time, skill, and cost needed for an operation. He tried to implement division of labor by using highly skilled laborers for high-skill jobs and low-skilled laborers for low-skill jobs, thus optimizing human resource and capital. He was a firm believer of the notion "right person for the right job". Charles Babbage compiled the findings from his factory visits into a book entitled "On the Economy of Machinery and Manufacturers" in 1832 (https://www.britannica.com/biography/Charles-Babbage, n.d.). His book emphasized the division and specialization of labor, scientific time and motion study for manufacturing a component, and organization of work for improving productivity. Charles Babbage is also famous for his contribution to evolution of computer and many regard him as a father of the modern computer.

2.7 Evolution of Production/Industrial Engineering in the Period Under Review

During First Industrial Revolution, the main focus was on the development of machines that could replace human power. Of course, the manpower was needed to operate the machines and carry out other assistive tasks. A lot of emphasis was put on machines and men in this period. The factory system of manufacturing emerged during this period, which necessitated the division of labor and curtailed the skills of an artisan as a side effect. The concept of mass production and standardization also developed in this period. It is often said that Industrial Engineering attempts to manage five "M"s— man, machine, material, method, and money. This period laid the foundation of Industrial Engineering.

3 PERIOD OF SECOND INDUSTRIAL REVOLUTION (1870–1914)

As a result of First Industrial Revolution, the factory system of production was developed. The period of the First Industrial Revolution is characterized by the development of steam engine, textile machineries, and machine tools. The pace of invention slowed down after about 1825. Although there were a number of industries, they were not properly organized and managed. In order to survive in the market, it was necessary to optimize the resources comprising man, material, and machine. Also, efforts were made to produce the goods of mass consumption, which necessitated mass production techniques. Development of efficient road and rail transportation system helped in the mass distribution of goods. The sewing machine was developed in 1844 and high volume canning and packaging devices were developed in mid-1880s. Soon, the Second Industrial Revolution started (Jensen, 1993; Liao et al., 2017). Although there is a difference of opinion among the historians regarding the exact duration of the Second Industrial Revolution, most of them consider it to be a period starting from 1870 to 1914, i.e., the beginning of First World War. Remarkable developments during the Second Industrial Revolution include the Bessemer process for steel production that almost halved the cost of production from early 1870 to late 1890, electrolytic refining process for the extraction of aluminum, production of plastics, and reduction in the cost of production of sugar and oil. However, the most remarkable achievement of this period is the development of scientific management and the formal recognition of Industrial Engineering. In this period, significant developments took place in science following reductionist approach, which is an effort to explain the phenomenon related to aggregates in terms of the parts by which the aggregates are formed (Andersen, 2001). This approach of scientific research induced industrial research also. Hence, it is no surprise that scientific management developed in this era. Significant contributions of some leading personalities of this time are presented hereunder.

3.1 Frederick Winslow Taylor (1856–1915)

Frederick Winslow Taylor, the famous American Mechanical Engineer, is often called the father of Industrial Engineering (Copley, 1923). He is considered as the pioneer of the "efficiency movement" and "scientific management." During his period, there was no proper management of the processes in the factory system and efficiency was not the best possible. Taylor studied, analyzed, and experimented with the nature of work in the factories and formulated some techniques to improve the production efficiency. He made extensive use of the concepts of division of labor and time and motion study in manufacturing. His technique of enhancing efficiency included assessing the standard time to perform a work and standardization of work process. His scientific approach led to better work methods replacing the previous method of "rule of thumb." Taylor's technique was named as "scientific management." The root of comparatively newer concepts like production planning and control and MRP (material requirements planning) can be traced back to Taylor's Shop Management methods (Wilson, 2016). Production planning is a plan for future production that requires proper knowledge of availability of materials, facilities, manpower, customer need, and demand. F.W. Taylor is best remembered for his efforts to evaluate the "efficiency" or "capability" of men and optimize it. There was no study and knowledge of human performance capability in quantitative manner until Taylor's scientific work. Taylor compiled his new concepts and ideas for higher industrial efficiency in his book entitled "Principles of Scientific Management" published in 1911. Some of the key principles of Taylor's scientific management are as follows:

• Division of a work into smaller parts and exploration of the best way of accomplishing each part
• Division of labor and also ensuring that right person does the right job
• Setting standard time for doing a work
• Providing proper training and financial incentives to the workers

Taylor's principles and practices, together with contributions from other followers, laid the foundation of Industrial Engineering. Taylor's idea of scientific management became very popular in America and industries reaped the benefits. Taylorism spread around the globe in the early part of 20th century. Industrialists in Britain, France, Vienna, and Russia followed Taylor's concepts for higher productivity. Consultants started implementing scientific management principles and training the employees in the industries. Implementation of scientific management resulted in organized factory system with better work culture among the workers. Thus, F.W. Taylor has an important place in the history of Production and Industrial Engineering for his scientific and methodical ideas and implementations.

F.W. Taylor also initiated first scientific study on metal cutting and had invaluable contribution toward metal cutting and tool life. Taylor and White developed a cutting tool material called high speed steel (that retains its hardness at elevated temperature) that caused about fourfold increase in the productivity (Dixit et al., 2017).

3.2 Henry Robinson Towne (1844–1924)

Henry Robinson Towne was an American Mechanical Engineer who has great contributions to the growth of Industrial and Production Engineering. He owned Yale and Towne Manufacturing Company and initiated mass production of electric hoist, testing machines, cranes, etc. Some of his key ideas are contained in a lecture that he delivered at Purdue University on February 24, 1905 (http://www.stamfordhistory.org/towne1905.htm, n.d.). H.R. Towne was a firm believer of the scientific management principles as laid down by his contemporary F.W. Taylor and tried to implement it in his company. He understood the significance of management techniques and emphasized the need to unite Production Engineering and Management to gain higher productivity. His views on higher productivity were reflected in the paper "The Engineer as Economist" published in Transactions of American Society of Mechanical Engineers (ASME) in 1886. Towne was elected as President of the ASME in 1888.

3.3 Henry Laurence Gantt (1861–1919)

Henry Laurence Gantt was an American Mechanical Engineer and management consultant who was a zealous follower of the Taylor's principles of scientific management. He was a colleague of F.W. Taylor in Midvale Steel and Bethlehem Steel and worked together for the application of scientific management principles in the factories. The famous "Gantt Chart" for activity scheduling was conceived by Henry Gantt in 1912, which demonstrated the proper schedule of the organizational activities (Herrmann, 2006). It is a systematic graphical representation of planning and scheduling of work, recording the progress of the work, and rescheduling if needed. Graphical scheduling tools can be traced back to 18th century. Joseph Priestley, William Blake, William Playfair, and Karol Adamiecki are some of the earlier contributors to the development of scheduling (Weaver, 2006). Gantt extended Taylor's work by inventing additional technique for measuring worker efficiency and productivity and designing "task and bonus" system for wage payment. He was of the opinion that the bonus paid to the managers should be proportional to how well they taught their workers to improve performance. Based on his opinions and inventions, Gantt published two books, "Work, Wages, and Profits" in 1916 and "Organizing for Work" in 1919.

3.4 Frank B. Gilbreth (1868–1924) and Lillian Gilbreth (1878–1972)

Frank Gilbreth and Lillian Gilbreth formed the famous innovative husband-wife team that revolutionized modern Industrial Engineering. They have immense contribution in shaping up Industrial Engineering to its modern stage. The couple followed in the footsteps of F.W. Taylor and gave concrete shape to Taylor's concept of "time study" in performing a work. The Gilbreths primarily followed the steps of identification, analysis, and measurement of basic sequences (or motions) involved in completing a work. Frank and Lillian Gilbreth firmly believed that there could be only one best and the most efficient way of doing a work. They conducted extensive studies on the method of doing a work regarding sequence of operations, time needed, and idle motion and time. Each task was divided into small elements, and essential elements were identified. Further, the optimization was carried out by eliminating the unnecessary motions and identifying the best sequence (motion) of these elements for getting the maximum possible productivity. They were pioneers in using motion pictures for motion study and analysis of work and the workers (http:// gilbrethnetwork.tripod.com/bio.html, n.d.). Frank Gilbreth was the first to propose that a nurse should help a surgeon by passing surgical instruments to reduce time and unwanted motion during surgery. He introduced various improved techniques for reducing unwanted motion in the activities of the army and bricklaying methods. Thus, the Gilbreths introduced "time and motion study" to identify the most productive and the most efficient method of executing a work. The Gilbreths tried to implement their rule "one best way of doing a work for higher efficiency" in their household activities. Their other innovative contributions toward scientific management, mass production, human factors and ergonomics, and psychology had revolutionary effect on Industrial Engineering, productivity, and management practice in an industry. The famous quote by Frank Gilbreth, "they come cheaper by the dozen" became a popular phrase for mass production and automation in the 20th century (http://www.up.ac.za/media/shared/404/ ZP_Files/Innovate%2009/Articles/the-roots-of-industrial-engineering_k-ruger.zp39685.pdf, n.d.).

Lillian Gilbreth was the first industrial psychologist to integrate psychology with industrial management and obtained a PhD degree on Psychology of Management in 1914. The concept of industrial management by considering the personal need of the workers, interpersonal relationships, and differences was initiated by Lillian Gilbreth. Her management techniques were very successful for improving productivity in an industry. She planted the seed of the human relation movement that emerged subsequently in the 1930s. Frank Gilbreth and Lillian Gilbreth are immortalized in the classic books "Cheaper by the Dozen" and "Belles on Their Toes" authored by two of the Gilbreths' children as a tribute to their parents.

3.5 Henry Ford (1863–1947)

Henry Ford, the famous American industrialist, will be always remembered for his contributions toward mass production by introducing assembly line technique in the automobile industry (Curcio, 2013). He was the founder of The Ford Motor Company where he tried to implement the Taylor's principles of scientific management for mass production of cars. Until that time, a car was considered as a luxury and only the rich could afford it. Henry Ford gifted the world the first "Ford Model T" in 1908, a car affordable to middle class people that revolutionized conveyance and transportation in early 20th century. The sale and popularity of Model T skyrocketed due to its lower price and easier driving system. Introduction of assembly line production technique in 1913 led to significant reduction in the assembly time of a car, enabling an enormous increase in production volume. This resulted in the reduction of production cost of cars. Sale volume increased and half of all cars in America were Ford Model T by 1918. Henry Ford initiated the trend of providing financial incentives to employees for increasing productivity. He was of the opinion that an employee had to be financially stable to extract the highest efficiency out of him. Providing incentives, a share in the profit, leisure time between work hours, and 5-day work-week were some of his ideas for getting the best service from his employees. Henry Ford's efforts for mass production of goods with lower price and higher wages for the employees had significant impact in the industries in the 20th century. His techniques were used around the globe, thus spreading his ideology, popularly known as "Fordism." He was a visionary with commitment who could think ahead of his time which led to many technical and business innovations.

The Mass production technology popularized by Henry Ford and his contemporaries spread around the globe after the First World War. The Ford Motor Company started the mass production of military tractors, trucks, and aircrafts for use in the First and Second World Wars. A new breed of Mechanical Engineers appeared in 1920s whose goal was to improve productivity in the industries and they were named as Production Engineers. As per suggestions of H.E. Honer, on 26th February, 1921, the Institution of Production Engineers was founded (Armytage, 1961). By middle of the 20th century, the United States had become the dominant industrial and economic power. The Ford Motor Company was manufacturing about one third of all the world's automobiles by 1930s. Mass production was diversely experimented in automobile industry, metal industry, naval industry, and electrical and chemical industries. Some of the other notwithstanding events of this period are as follows:

- Formulation of "Queueing theory" in 1909 by Agner Krarup Erlang (1878–1929), the famous mathematician from Denmark, which was later applied in statistical quality control methods
- Formulation of "Markov chain" and "Markov process" by a contemporary Russian statistician, Andrei Markov (1856–1922), which had wide applications in queues, inventory control, and stochastic systems
- Development of the concepts of administrative management approach and systematic practice by H. Fayol, M. Weber, and C. Barnard (Nadler, 1992).
- Development of statistical analysis techniques for checking quality of a product and laying the foundation for Statistical Quality Control (SQC) by Walter Shewhart in 1924 (Walter Shewhart also developed the Shewhart Cycle with four steps: Plan, Do, Study, and Act, commonly referred to as the PDSA cycle that led to total quality improvement.)

3.6 Evolution of Production/Industrial Engineering in the Period Under Review

It is observed that during this period, a lot of developments took place in formulating proper methods for carrying out a task. Management techniques were developed as a science and physics and applied by the workers in executing their works. On the inventory management front, an economic order quantity formula (EOQ) was published by F.W. Harris (Roach, 2005). EOQ estimates the order quantity for minimizing the total cost. The disciplines of Industrial and Production Engineering took birth in this period. Significant developments took place in the materials side also. Bessemer process for steel production and high speed steels was developed in this period. Use of electrical power in industries became prevalent. Thus, man, machine, material, and method gained prominence in this period. A historical survey of the production methods, their control, and operation management techniques used during early 19th century to mid-20th century can be found in Mckay (2003).

4 PERIOD OF HUMAN RELATION MOVEMENT (C.1930–1950)

The seed of human relation movement was planted in the 1930s from the eventual realization that the human factor is a crucial element for higher productivity. In the early stage, industrial management considering worker relationships, personal need of the workers, and incentives was initiated by the visionaries like Lillian Gilbreth and Henry Ford which eventually led to the human relation movement in 1930s. Elton Mayo, Douglas McGregor, and Abraham Maslow are some of the pioneers who had significant contributions to the human relation movement. The movement emphasized the significance of human factors in industry and its importance for production. The key points identified as human factors are interpersonal relationship of the workers, their motivation and response, human behavioral aspects, teamwork, incentives, mental/physical stress related to work, etc. "Personnel Management" was considered as a subject of research and building and maintaining an efficient and effective work-team was the responsibility of a production manager. In this period, industries appreciated the importance of human management apart from the management of technology.

4.1 Elton Mayo (1880–1949)

Elton Mayo, the Australian psychologist, was famous for his contributions to the human relation movement in the 1930s and the concepts of behavioral management through "Hawthorne studies." He took keen interest in human psychology and behavior within an industry and conducted studies associating the same with productivity. Mayo's research was mainly based on the experiments he conducted among the workers in Hawthorne plant, Illinois, in the late 1920s and early 1930s (https://mbsportal.bl.uk/taster/subjareas/busmanhist/mgmtthinkers/mayo.aspx, n.d.). He was of the view that relationship among the workers of an organization is a crucial factor for industrial organizations. Mayo applied his knowledge of psychology to formulate new ideas in the fields of industrial research, business management, and industrial sociology. His ideology and research formed the stepping stones to the human relations movement. Elton Mayo offered counseling to the soldiers of First World War to recover from the stresses of war. He also served as a professor of industrial research in the Harvard Business School, USA, in 1926. Elton Mayo's ideology is reflected in his book "The Human Problems of an Industrialized Civilization" published in

1933 (Mayo, 1933). His studies in industrial research had a significant impact on associating industrial and organizational psychology with enhanced productivity in an industry.

4.2 Abraham Maslow (1908–1970)

Abraham Maslow of America was another famous psychologist who shaped up the human relation movement initiated by Elton Mayo. Maslow was famous for his motivation studies and theory of hierarchy of needs. According to Maslow's theory of hierarchy presented in 1943, people have five basic needs that can be ranked in the following hierarchical order: (1) physiological needs (can be termed as the need for survival), (2) need for safety and security, (3) craving for love and belongingness, (4) need for self-esteem, and (5) self-actualization. He was of the view that, as long as one works to satisfy these needs, one gives the best performance, thus enhancing efficiency, growth, and productivity. The need of monetary incentives, sense of belonging, and social acceptance of the workers in an industry was emphasized by Maslow for higher output. After the first four basic needs are satisfied, every individual desires for "self-actualization," i.e., achieving a level where he can use his potential or capability to the fullest. It is a type of spiritual need for which an individual works without any physical reward. It is like an artist wants to make an artifact for self-satisfaction. Maslow's ideas of behavioral management influenced the managerial attitude toward workers in that period. His notions were popularly known as positive psychology and constituted valuable inputs to the human relation movement that reached its peak around 1960. Maslow's theory is relevant today also. For example, a recent survey showed that salary had only a 20% impact on job satisfaction (Sadri and Bowen, 2011). The management should try to satisfy other higher needs of the employee.

4.3 Douglas McGregor (1906–1964)

Douglas McGregor, the American Mechanical Engineer, was a contemporary of Abraham Maslow and shared the same passion for human relations and behavioral management study. In 1937, he founded a department of industrial relations at MIT. He was famous for his Theory X and Theory Y which best represents the essence of the human relations approach. According to Theory X, the employees dislike and even avoid work and will find themselves in a difficult-to-control environment, while according to Theory Y, the activities in an organization are a source of personal satisfaction (Simionel, 2011). Theory X is represented by the bad persons and Theory Y by good persons. It is possible to motivate people to change from X to Y type. Theory Y is more positive and appropriate to be followed in an industry for dual benefits of the workers and the industry. McGregor published his theories in the book "The Human Side of Enterprise" in 1960. According to McGregor, management by directing and controlling the workers (whose basic needs are not satisfied) fails as it does not provide enough motivation. Workers feel deprived and become passive. He believed that an abundant resource of creative human energy could be tapped with proper motivation and appropriate working conditions in an industry. McGregor recognized physiological, safety, social, egoistic, and self-fulfillment needs as predominant in an individual (McGregor, 1961). A worker can be motivated to give his best performance by satisfying his basic needs.

4.4 Evolution of Production/Industrial Engineering in the Period Under Review

Human relations and behavioral management studies gave valuable insights into worker's psychology and interpersonal affairs in an organization in this period. The Hawthorne experiments as well as the concepts of hierarchy and satisfaction of need showed how productivity is significantly increased by enthusiastic participation of the workers. With these developments, Production and Industrial Engineering flourished until and after the Second World War. Although mass production techniques were used in industries, they practically gained momentum during the outbreak of Second World War. As restrictions were imposed on imports and exports, the industries were forced to produce necessary articles for war in large quantities using mass production technology. The first jet aircraft Heinkel and helicopters were developed in Germany in 1939 and 1941, respectively, during Second World War (Garrison, 1999). A new topic known as "ergonomics" was introduced in 1940s, which dealt with the issues related to the human resources (workers). In ergonomics, statistical tools and human physiological data are used to reduce health and safety risks of humans using mechanical products. Gradually, industrial management deviated toward quantitative management approach with the development of mathematical and statistical tools. The first seed of quality movement was planted with the implementation of scientific management and development of statistical analysis and control theories that were later applied in industries. In 1948, American Institute of Industrial Engineers was formed to provide a common platform to the Industrial Engineers. However, in 1981, the word "American" was removed.

Circumstances during and after Second World War provided a lot of impetus to the development of industrial engineering. Countries recognized the scarcity of resources including manpower. They also needed to enhance production of arms and related goods. A need was felt to optimize the material resources and human effort. Hence, the areas like operations research, ergonomics, and work study got importance.

The period from 1950 to 1970 can be considered a period of intense research in the field of electronics, computer science, and computational techniques (Dixit et al., 2017). During this period, John D.C. Little presented Little's law famous in queuing theory (Little, 1961). The first numerical control machine tool was developed at Massachusetts Institute of Technology in 1952. In 1950s, transistors were used in several electronic devices. G.W.A. Drummer proposed the concept of integrated circuits in 1952, which revolutionized electronics industry. Texas Instruments developed the first commercial integrated circuit in 1958. First computer to make use of integrated circuits was produced in 1968. The first handheld calculator was produced in 1967. All these developments caused Third Industrial Revolution.

5 PERIOD OF THIRD INDUSTRIAL REVOLUTION (1970−2000)

Several researchers consider the period starting from 1970s and up to 1990s as the period of Third Industrial Revolution (Jensen, 1993; Liao et al., 2017). This period is characterized by the use of electronics and information technology in manufacturing. The fields of computers and electronics were revolutionized by the works of the giants like Michael Dell, Gordon Moore, Rod Canion, and Steve Jobs. Michael Dell is the founder of Dell Technologies and has turned it into one of the world's leading computer companies. Same can be said for Steve Jobs, founder of Apple Company, and Rod Canion, founder of Compaq Company. Gordon Moore is the founder of Intel Corporation and is credited for introducing Moore's law (Moore, 1998). Industrial Engineers contributed in automating the manufacturing facilities and producing quality products. In this section, brief notes on the developments of automation and quality movement are presented. While the automation was pioneered by Electronics and Computer Engineers, Industrial Engineers and Management gurus contributed toward enhancing the quality of the products.

5.1 Automation

In c.1720, the first programmable loom using punch cards was developed in France (Deb, 1994). From the middle of 20th century, there was a growing need for automation in the industry to meet the workload due to increase in production volume and need to enhance quality. Automation of production process is advantageous as repetitive and routine tasks could be automated, thus saving time and labor as well as improving productivity and quality. Specific areas where automation was implemented were machining, welding, transportation, assembly, inspection, quality control, and packaging. In a production industry, automation mainly refers to the application of automated guided vehicles (AGV), computer numerical control (CNC) machines, and robots. A.M. Barrett, Jr., introduced the first AGV in 1954 which was a modified towing truck. The combination of automatic storage and retrieval system (ASRS) and related software led to development of flexible manufacturing system (FMS), which was installed in a number of American firms in 1970s and 1980s (O'Grady and Menon, 1986).

Synergistic integration of Electronics and Mechanical Engineering led to conception of Mechatronics in 1960s, which had significant effect on the products and production technologies. Mechatronics is extensively used in robotics, automation, intelligent motion control, flexible manufacturing systems (FMS), and CNC machine. Mechatronics set the stepping stones for Third Industrial Revolution. Many primarily mechanical products were replaced by electronic products, for example, watches.

Role of Mechatronics was instrumental in emergence of Robotics in 1960s and 1970s. Planet Corporation is credited for inventing the first commercial robot in 1959. Use of robots has become essential for tasks that are repetitive, dangerous, and cannot be accessed by human beings. Robots are extensively used in production and manufacturing for welding, machining, assembly work, as well as moving materials, parts, and tools.

There was wide use of Numerical Control (NC) technology in the production industry for mass production in 1970s. Introduction of computer control to NC machines improved its performance and made it Computer Numerical Control machines. MIT led the CNC revolution by developing the first CNC machine tool in 1957 and it spread to United States and Germany followed by Japan and other countries across the globe. CNC technology has led to enhancement in quality and variety in products with less human intervention. Inception of computer-aided design (CAD), computer-aided manufacturing (CAM), and computer-integrated manufacturing (CIM) was the result of using CNC technology. Implementation of these systems enables faster design, production, and testing, thus bringing revolution to the way goods are produced.

In the early days of automation, group technology was used to geometrically categorize similar products. One particular CNC machine could produce slightly different parts with minor changes in tools and fixtures, thus enabling flexible manufacturing system (FMS). FMS became increasingly popular in the United States, Europe, and Japan during the mid-1970s. In the postindustrial era, trend was to adopt integrated and parallel working approach for automation rather than prevailing sequential approach of automation and integration (Vonderembse et al., 1997). Flexibility is very important in this age of ever-changing customer's choice. Use of FMS and RMS (Reconfigurable manufacturing system) is the call of the day to adapt to these changes. Rapid Prototyping (RP) is a promising technology in the recent times that can directly fabricate a component from 3D CAD model. RP is gaining prominence for fabricating customized products like cosmetic dentistry, hearing aids, and human body parts replacements. Use of artificial intelligence (AI) and soft computing techniques in production process is another achievement in this period. AI is being successfully applied in different areas, viz., design, optimization, and manufacturing, all control activities as well as simulation.

With the initiation of Internet and World Wide Web in 1990s, globalization effect sets in within the industries. Application of these new technologies imparted more flexibility and allowed customization. The new challenge in the competitive global market is to produce goods of best quality at optimum cost at the shortest time. Customers' tastes and preferences are ever changing and they prefer both variety and quality in products. Adaptation to the changing scenario is a crucial factor in this era of lean and agile manufacturing. Taiichi Ohno (1912–1990) is popularly known as the father of Toyota Production System, which came to be known as lean manufacturing system in the United States. He was born in Dalian, China, and joined Toyota Automatic Loom Works around the World War II period (https://in.kaizen.com/blog/post/2013/10/17/taiichi-ohnos-contribution.html, n.d.). He joined the car manufacturing company of Toyota at the end of the World War II, at the time when Japan was much behind America in automobile production. Ohno emphasized on the elimination of waste and is pioneer in developing popular Toyota Production System (TPS). He developed Kanban system of Just-In-Time inventory control. Sakichi Toyoda, founder of Toyota Group, and his son, Kiichiro Toyoda, also have a lot of contribution in developing TPS. Lean and agile manufacturing systems started in 1990s with a goal to minimize waste and adaptation to changes. It has become essential for a production industry to use cellular manufacturing system, FMS, and Reconfigurable Manufacturing System (RMS) approaches to cater to these needs. Cellular manufacturing is a subset of Group Technology, which originated in U.S.S.R. in 1940s. "Group Technology can be defined as bringing together and organizing (grouping) common concepts, principles, problems and tasks (technology) to improve productivity" (Greene and Sadowski, 1984). S.P. Mitrofanov of U.S.S.R. published a book entitled "The Scientific Principles of Group Technology" in 1958. In a production company producing a number of components, components can be classified into groups. The components in a group may have similarity and can be manufactured in the same cells. Industry can have as many cells as the groups of the components, such that a group can be assigned to a proper cell. The cell contains appropriate machines for producing all components of the group. This approach is called cellular manufacturing. Professor H. Optiz of Aachen Technical University, Germany, developed a part classification and coding system in 1960s. John L. Burbidge developed a Production Flow Analysis (PFA) method for configuring a cellular manufacturing system in early 1970s. Process planning plays an important role in bridging the gap between design and manufacturing of a part. It basically converts the design requirements of a part into manufacturing instructions. Process planning involves proper selection of raw material, machining processes and their sequence, process parameters (cutting velocity, feed, depth of cut, etc.), machine tool, cutting tools, and fixtures so that a part can be manufactured with good quality at optimum cost. It has direct influence on product quality, cost, machining efficiency, and so on. The inception of computer-aided process planning (CAPP) dates back to 1966 with the formation of a CAPP working group in Paris by R. Weill (Hoda and Maraghy, 1993).

With the emergence of Java and Web technologies, web-based production systems have been designed to share information over Internet. In this era of e-manufacturing, it is possible to perform designing, fabrication, and delivery of a component at different places across the globe. Moreover, global logistics has imparted more flexibility to production systems by providing easier transportation of raw materials and finished goods and other services like networking, packaging, warehousing, optimal routing, etc. Thus, in the 21st century, cutting-edge technology and modern scientific management has elevated Production and Industrial Engineering to a new height. However, it is difficult to mention only a few names as the pioneer of this movement; it is a collective effort.

5.2 Quality Movement

Managing the Production and Industrial Engineering by quantitative approach was initiated in 1940s and continued to 1960s, resulting in the development of mathematical and statistical tools like Decision Support Systems, Linear Programming, Operation Research technique, Simplex method and optimization, Project Evaluation Review Technique, Critical Path Method, and Material Requirement Planning, which were widely applied in the industries. Efficiency and productivity were given more importance in this approach than customer satisfaction. The famous mathematician George B. Dantzig is credited for introducing simplex method for efficient solution of linear programming problems (Gill et al., 2008). Some of the pioneers, worth mentioning in the field of linear programming, are Leonid Kantorovich, Frank Lauren Hitchcock, and John von Neumann. However, gradually manufacturers realized that products and services should be produced to meet the need of the customer. Quality became an important factor in the customer's need. Industries realized that focusing on quality improves overall productivity and reduces costs. Quality rather than quantity was emphasized and the trend was to produce fewer and better products. Mass customization was first introduced by Ford Motor Company in 1964 by allowing customers to select their preferences in buying cars. A host of new concepts like just in time (JIT), total quality management (TQM), lean and agile manufacturing, as well as concurrent engineering were developed during this period. All these concepts basically strive for continuous improvement and minimum wastage in an industry. JIT, developed in Japan to meet the crisis of shortage of manpower and resources after the Second World War, aims to eliminate wastage of resources (raw materials, manpower, labor, time), work-in-progress, scrap, and inventory. JIT promotes lean manufacturing rather than mass manufacturing. The concept of "quality at the source" is encouraged in lean manufacturing by adding value to the product at different stages of production. Concurrent engineering encourages active participation of all concerned in parallel progress of design, manufacturing, quality control, and marketing. The concept of TQM gained momentum in 1970s. TQM is the name given to a systemic approach for management of organizational quality. Modern management theories like "Kaizen" and "Kanban" (meaning "continuous improvement" and "change for better", respectively) were first implemented in Japan. These ideas were spread around the world and industries reaped their benefits. Thus, quality revolution reached its peak around 1970s and continued till date. Contributions of Deming, Juran, Taguchi, Crosby, and many others are indispensable for the quality movement (Logothetis, 1992). Their contributions are presented in this section.

5.2.1 W. Edwards Deming (1900–1993)

W. Edwards Deming, the American statistician and quality expert, is revered across the globe for his invaluable contributions to the quality movement. Deming drew inspiration from Walter Shewhart who planted the seed of "total quality management (TQM)" as early as 1924 by developing statistical analysis techniques for checking quality of a product. Deming's work in Japan in 1950s after Second World War and in America in 1980s had revolutionary effect in implementing TQM in production and industrial fields. Deming taught Japanese engineers statistical methods for analysis and control of quality. It can be considered the foundation of TQM. His contribution is indispensable for Japan's reputation for innovative and high-quality products. Deming's philosophy for continuous improvement was the "Plan-Do-Check-Act" cycle based on Walter Shewhart's work. Deming is best known for his "14 Points," the key principles that an organization has to adopt for customer satisfaction through quality (Deming, 1986). Some other famous concepts formulated by Deming for better quality and continuous improvement are "System of Profound Knowledge," "Prevention by Process Improvement," and "Chain Reaction for Quality Improvement." Deming put the major responsibility for quality on management. He was against performance appraisal and believed in positive motivation and scientific approach to manufacturing (Logothetis, 1992).

5.2.2 Joseph M. Juran (1904–2008)

Joseph M. Juran was an American engineer and management consultant who contributed significantly toward quality movement. He was contemporary of Edwards Deming and is considered one of the three experts (Deming, Juran, and Crosby) of quality management. Juran authored several books on quality control and total quality management. According to Juran, quality meant fitness for use. In the initial stage, he mainly focused on managing the quality through application of Pareto principle, control charts, sampling, and inspection. He tried to associate the human factor to quality management. The important strategies proposed by Juran for quality management are: "Spiral of Progress in Quality," "Breakthrough Sequence," "Project-by-Project Approach," "Juran Trilogy," and "Vital Few and Trivial Many." "Juran Trilogy" is a systematic approach consisting of three steps: quality planning, quality control, and quality improvement for attaining and managing quality (Juran, 1986).

5.2.3 Genichi Taguchi (1924–2012)

Genichi Taguchi, the famous Japanese engineer and statistician, is best remembered for his statistical methods for improving quality. He had the privilege of learning the statistical quality control as introduced by Walter Shewhart and Deming in Japan and applies the same for improving quality of a product. "Taguchi method," developed by him based on statistics, is best suited for improving the quality of design. Some of his greatest contributions to Industrial Engineering and quality management are: "use of design of experiments (DOE)," "Taguchi loss function," and "Taguchi method."

5.2.4 Philip B. Crosby (1926–2001)

Philip B. Crosby was a famous quality expert and consultant who had significant contributions in the domain of quality and quality control. He is best known for his concept of "zero defects" that originated in the United States in 1960s and spread worldwide. The implementation of "zero defects" paved the way for quality improvement in many companies. According to Crosby, quality means "conformance to requirement" and it should be defined in measurable terms. The cost of nonconformance is the cost of not doing it right the first time (https://mbsportal.bl.uk/taster/subjareas/busmanhist/mgmtthinkers/crosby.aspx, n.d.). Prevention rather than correction was the foundation for Crosby's work. Some of the other theories put forward by Crosby, which had extensive effect on quality movement, are "Do It Right the First Time," "Zero Defects," "Four Absolutes of Quality," "Prevention Process," "Quality Vaccine," and "Six C's." He authored several influential books starting with the best-seller "Quality is Free" in 1979 that produced great impact on the quality movement in the United States and Europe.

5.3 Evolution of Production/Industrial Engineering in the Period Under Review

In this period, Industrial Engineering focused toward facilitating automation, enhancing the quality and productivity, using information technology in manufacturing, and practicing lean and agile manufacturing. The quality control methods developed during this period are still prevailing in the industry. For example, application of Six Sigma method of quality control solved tolerance-related errors and earned huge profit for a company producing automotive parts (Antony et al., 2012). This period is characterized by putting emphasis on diversification of management strategies from industry to other sectors (Keulen and Kroeze, 2014). The Industrial Engineers also started working in service sectors. Industrial Engineering and Management expanded its domain to banking, insurance, travel, and many similar organizations. It got involved in managing all 5 'M's including money.

6 PAST, PRESENT, AND FUTURE OF PRODUCTION/INDUSTRIAL ENGINEERING

Table 1 summarizes the evolution of Production and Industrial Engineering. Production and Industrial Engineering has reached the stage of e-manufacturing at the click of a button from the early stage of homemade

Table 1 Evolution of Industrial Engineering

Period	Main features	Main contributors
1750 – 1850	Division of labor, replacing muscle by machine, concept of interchangeability	Adam Smith, Eli Whitney, Samual Bentham, Marc Brunel, Henry Maudslay and James Watt, Charles Babbage
1851 – 1930	Method study, distinct identity of industrial/production engineering, use of electrical power, steel production, SQC, queuing theory	Taylor, Gilbreth, Towne, Ford, Gantt, Shewhart, Erlang, Honer, Diemer
1931 – 1960	Ergonomics, quality management, operations research	Maslow, Mayo, McGregor, Deming, Juran, Dantzig, Walker, Kelley, Hamilton, U.S. Navy
1961 – 2010	Automation, quality, supply chain management, lean and agile manufacturing, six sigma, web-based production, cloud computing, and manufacturing	Taguchi, Ohno, Morley, Jobs, Gates
2010 onward	Industry 4.0, Internet of things, sustainable manufacturing, big data and analytics, digital manufacturing	German Federal Ministry of Education and Research, European Union, companies like Google

products. Natural urge of human being for continuous improvement, change in their needs, cutting-edge technology, stiff competition, and globalization acted as fuels to shape Production and Industrial Engineering to its state-of-the-art form. Some of the areas that have tremendous potential in the present as well as future are briefly touched hereunder.

The most important and crucial field in the present time is sustainable manufacturing for attending to the issues of the environment and global warming. In-depth research in environmentally friendly methods of manufacturing is to be widely explored. There is a worldwide drive for extraction of green and renewable energy from wind, water, the sun, and biomass to meet the energy crisis and protect the environment. Alternative energy and environmental engineering are expected to gain prime importance in the near future for sustainable manufacturing. On the technological front, 3D printing (an additive manufacturing method) is a promising technology where parts can be manufactured layer by layer from 3D CAD model. It is likely to have a profound impact on production of goods due to its flexibility and scope of customization. It is expected that additive manufacturing will be instrumental in initiating the Fourth Industrial Revolution, "Industry 4.0," with the concept of "Digital Transformation of Industries" (http://phys.org/news/2016-01-industry-additive.html, n.d.). The latest control strategies like ETC (event-triggered-control) and STC (self-triggered-control) will accelerate the digital transformation of the manufacturing methods (Dotoli et al., 2017). Many believe that Fourth Industrial Revolution has already started with the use of cloud computing and Internet of Things. E-commerce and e-manufacturing will revolutionize the Production and Industrial Engineering in the near future. Some of the cutting-edge fields that are likely to dominate for the next few decades are virtual prototyping, nanotechnology, motion simulation, more advances in robotics and mechatronics, automation, machine vision, latest computer technology and electronics, green technology, etc. Emphasis is to be placed on global collaboration, networking and management, knowledge sharing, and integration. Future of Production and Industrial Engineering has extreme potential and it will continue to set new milestones and explore new heights with its pioneering works.

Future developments in industrial sector can be covered under a common heading—Industry 4.0 (Zhong et al., 2017). In 2013, German government launched its Industry 4.0 plan. European Union launched Horizon 2020, a research and innovation program toward Industry 4.0. In the United States, GE introduced the concept of Industrial Internet of Things emphasizing the use of cyber-physical system in near future. In 2015, Japan started its Industrial Value Chain Initiative, which is a counterpart of Germany's Industry 4.0. China started Made in China 2025. India has started Make in India initiative in September 2014. Digital manufacturing will pay a big role in it.

7 CONCLUSION

A brief history of Production and Industrial Engineering is presented in this paper through the contributions of the legendary personalities. Significant developments of Production and Industrial Engineering during its evolution are discussed with a perspective of the new technologies and their possible applications in the future. Modern trends in the further development of Production and Industrial Engineering are reflected and its importance in the solution of global problems in the 21st century is emphasized. It is observed that Industrial Engineering is adapting itself to changes in the technology. It started around First Industrial Revolution with a focus on mass production, mechanization, and division of labor. Later on, attempts were made to optimize man, material, machine, method, and even money. With Fourth Industrial Revolution known as Industry 4.0, Industrial Engineering is attempting to make the best use of digitization and automation and is expanding its scope to nonengineering sectors as well. It is observed that the external form of industrial engineering has been changing according to prevalent technology and needs of the society; however, in essence, always the focus has been on optimizing the resources in the presence of various technological, social, and political constraints and in increasing the value to cost ratio.

IIE Council of Fellows has identified 8 grand challenges for engineers in 2007 (Askin, 2009). These are as follows: 1. Reengineering healthcare delivery, 2. Creating a technology-oriented culture, 3. Engineering a sustainable society by 2100, 4. Developing better decision-making tools in a dynamic world, 5. Mitigating and responding to disaster, 6. Point-of-use manufacture, 7. Infrastructure construction, and 8. Safe, available, and affordable food and water. After about a decade, these challenges still remain. Information technology complemented by distributed implementation of hardware technology can cope up with these challenges. The role of industrial and production engineers will be more important than ever.

REFERENCES

Andersen, H., 2001. The history of reductionism versus holistic approaches to scientific research. Endeavour 25 (4), 153–156.

Antony, J., Gijo, E.V., Childe, S.J., 2012. Case study in Six Sigma methodology: manufacturing quality improvement and guidance for managers. Prod. Plan. Control 23 (8), 624–640.

Armytage, W.H.G., 1961. A Social History of Engineering. The MIT Press, Massachusetts.

Askin, R.G., 2009. Grand challenges for industrial engineering in the 21st century. IIE Annual Conference. Proceedings, Norcross. 6 pages.

Baofu, P., 2009. The Future of Post-Human Engineering: A Preface to a New Theory of Technology. Cambridge Scholars Publishing, New Castle upon Tyne.

Burstall, A.F., 1963. A History of Mechanical Engineering. Faber and Faber, London.

Bythell, D., 1983. Cottage industry and the factory system. History Today, April, 1983.

Capra, F., Luisi, P.L., 2014. The Systems View of Life: A Unifying Vision. Cambridge University Press.

Copley, F.B., 1923. Frederick W. Taylor, Father of Scientific Management. 2015 reprint, Facsimile Publisher, London.

Crump, T., 2007. The Age of Steam: The Power that Drove the Industrial Revolution. Robinson, London.

Curcio, V., 2013. Henry Ford. Oxford University Press, New York.

Deb, S.R., 1994. Robotics Technology and Flexible Automation. Tata McGraw-Hill Publishing Company, New Delhi.

Deming, W.E., 1986. Out of the Crisis. MIT Press.

Dixit, U.S., Hazarika, M., Davim, J.P., 2017. A Brief History of Mechanical Engineering. Springer, Switzerland.

Dotoli, M., Fay, A., Miśkowicz, M., Seatzu, C., 2017. Advanced control in factory automation: a survey. Int. J. Prod. Res. 55 (5), 1243–1259.

Garrison, E., 1999. A History of Engineering and Technology–Artful Methods. CRC Press, Florida.

Gill, P.E., Murray, W., Saunders, M.A., Tomlin, J.A., Wright, M.H., 2008. George B. Dantzig and systems optimization. Discret. Optim. 5, 151–158.

Greene, T.J., Sadowski, R.P., 1984. A review of cellular manufacturing assumptions, advantages and design techniques. J. Oper. Manag. 4 (2), 85–97.

Herrmann, J.W., 2006. A history of production scheduling. In: Handbook of Production Scheduling. Springer, New York, pp. 1–22.

Hoda, A., Maraghy, E.I., 1993. Evolution and future perspectives of CAPP. Ann. CIRP 42 (2), 739–751.

Jensen, M.C., 1993. The modern industrial revolution, exit, and the failure of internal control systems. J. Financ. XLVIII (3), 831–880.

Juran, J.M., 1986. The quality trilogy: a universal approach to managing for quality. Qual. Prog. 19 (8), 19–24.

Keulen, S., Kroeze, R., 2014. Introduction: the era of management: a historical perspective on twentieth-century management. Manag. Organ. Hist. 9 (4), 321–335.

Liao, Y., Deschamps, F., Loures, E.d.F.R., Ramos, L.F.P., 2017. Past, present and future of Industry 4.0—a systematic literature review and research agenda proposal. Int. J. Prod. Res. 55 (12), 3609–3629.

Little, J.D.C., 1961. A proof for the queuing formula: $L = \lambda W$. Oper. Res. 9 (3), 383–387.

Logothetis, N., 1992. Managing for Total Quality: From Deming to Taguchi and SPC. Prentice-Hall of India, New Delhi.

Martin-Vega, L.A., 2004. The purpose and evolution of Industrial Engineering. In: Zandin, K.B. (Ed.), Maynard's Industrial Engineering Handbook. McGraw-Hill, New York, pp. 1.3–1.9.

Mayo, E., 1933. The Human Problems of an Industrial Civilization. Harvard, Cambridge, MA.

McGregor, D., 1961. The human side of enterprise. Wiley, New York.

Mckay, K.N., 2003. Historical survey of manufacturing control practices from a production research perspective. Int. J. Prod. Res. 41 (3), 411–426.

McNiel, I., 1990. Introduction: basic tools, devices and mechanism. In: McNiel, I. (Ed.), An Encyclopedia of the History of Technology. Routledge, London, pp. 1–43.

Millerson, G., 1964. The Qualifying Associations: A Study in Professionalization. Routledge, London.

Mokyr, J., 1998. The second industrial revolution. 1870 – 1914, In: Castronono, V. (Ed.), Storia dell'economia Mondial. Laterza, Rome.

Moore, G.E., 1998. Cramming More Components onto Integrated Circuits. Proc. IEEE 86 (1), 82–85.

Nadler, G., 1992. The Role and Scope of Industrial Engineering, Handbook of Industrial Engineering. Wiley, New York, pp. 3–27.

Nanda, J.K., 2006. Management Thought. Sarup & Sons, New Delhi.

O'Grady, P.J., Menon, U., 1986. A concise review of flexible manufacturing systems and FMS literature. Comput. Ind. 7, 155–167.

Roach, B., 2005. Origin of the economic order quantity formula; transcription of transformation? Manag. Decis. 43 (9), 1262–1268.

Robert, V.B., 1989. Lincoln and the Tools of War. University of Illinois Press, Chicago.

Sadri, G., Bowen, R.C., 2011. Meeting employee requirements. Ind. Eng. 43 (10), 44–48.

Simionel (Sauciuc), A.-B, 2011. Douglas McGregor—theory X and theory Y. Rev. Manage. Econ. Eng. 10 (3), 229–234.

Vonderembse, M.A., Raghunathan, T.S., Subba Rao, S., 1997. A postindustrial paradigm: to integrate and automate manufacturing. Int. J. Prod. Res. 35 (9), 2579–2600.

Weaver, P., 2006. A Brief History of Scheduling. myPrimavera Conference. Hyatt, Canberra.

Wilson, J.M., 2016. The origin of material requirements planning in Frederick W. Taylor's planning office. Int. J. Prod. Res. 54 (5), 1535–1553.

Zhong, R.Y., Xu, X., Klotz, E., Newman, S.T., 2017. Intelligent manufacturing in the context of industry 4.0: a review. Engineering 3 (5), 616–630.

gilbrethnetwork   http://gilbrethnetwork.tripod.com/bio.html,   accessed   on   May   15, 2017 n.d.

phys.org http://phys.org/news/2016-01-industry-additive.html, accessed on May 17, 2017 n.d.

ice http://www.ice.org.uk, accessed on April 21, 2017.n.d.

iise http://www.iise.org/Home/, accessed on May 16, 2017.n.d.

imeche http://www.imeche.org/, accessed on April 25, 2017.n.d.

stamfordhistory http://www.stamfordhistory.org/towne1905.htm, accessed on May 15, 2017.n.d.

theiet http://www.theiet.org/resources/library/archives/research/guides-iet.cfm, accessed on April 25, 2017.n.d.

kruger   http://www.up.ac.za/media/shared/404/ZP_Files/Innovate%2009/Articles/the-roots-of-industrial-engineering_kruger.zp39685.pdf, accessed on May 15, 2017.n.d.

Kaizen https://in.kaizen.com/blog/post/2013/10/17/taiichi-ohnos-contribution.html, accessed on January 5, 2018. n.d.

mbsportal https://mbsportal.bl.uk/taster/subjareas/busmanhist/mgmtthinkers/crosby.aspx, accessed on May 17, 2017. n.d.

mayo https://mbsportal.bl.uk/taster/subjareas/busmanhist/mgmtthinkers/mayo.aspx, accessed on May 15, 2017. n.d.

Charles-Babbage https://www.britannica.com/biography/Charles-Babbage, accessed on May 7, 2017. n.d.

Eli-Whitney https://www.britannica.com/biography/Eli-Whitney, accessed on May 7, 2017. n.d.

Samuel-Bentham https://www.britannica.com/biography/Samuel-Bentham, accessed on May 4, 2017. n.d.

ieee https://www.ieee.org/index.html, accessed on November 9, 2017.n.d.

FURTHER READING

proquest https://search.proquest.com/docview/192457496?accountid=16262, accessed on December 5, 2017. n.d.

CHAPTER 2

Manufacturing Engineering Education—Indian Perspective

Prasanta Sahoo, Suman Kalyan Das
Department of Mechanical Engineering, Jadavpur University, Kolkata, India

Abbreviations

AICTE	All India Council for Technical Education
CSIR	Council of Scientific & Industrial Research
DRDO	Defense Research and Development Organization
DST	Department of Science and Technology
GDP	Gross Domestic Product
GIAN	Global Initiative of Academic Networks
IISc	Indian Institute of Science
IIT	Indian Institute of Technology
ISRO	Indian Space Research Organisation
ITI	Industrial Training Institute
JEE	Joint Entrance Examination
MHRD	Ministry of Human Resource and Development
MOOCs	Massive Open Online Courses
NAAC	National Assessment and Accreditation Council
NBA	National Board of Accreditation
NIRF	National Institute Ranking Framework
NIT	National Institute of Technology
NPIU	National Project Implementation Unit
NSDC	National Skill Development Corporation
OBE	outcome-based education
PG	Postgraduate
TEQIP	Technical Education Quality Improvement Program
UG	Undergraduate
UGC	University Grants Commission

1 INTRODUCTION

In the past, often manufacturing has not been considered as specific task of University-trained engineers (Peters, 1989). Manufacturing was mainly the job of higher technicians and foremen who were quite successful at it and production engineers used to supervise them and look after the overall

Manufacturing Engineering Education
https://doi.org/10.1016/B978-0-08-101247-5.00002-2

management mechanism. With time, in order to increase the efficiency of manufacturing, design, production, marketing, and management were integrated into a single system. This required the engineers to have an overall view of the manufacturing process as well as to have the capability to understand and optimize every technical component. Today, they also oversee the running, maintenance, and improvement of the production or manufacturing unit. It is the manufacturing engineer who is accountable for the efficient production of quality products with minimum cost.

From the ancient Bharat to a modern nation, India has always been an abode of higher education and learning. In ancient times, Nalanda, Vikramsila, and Taxila Universities were prominent places of higher education attracting students not only from all over the country, but from far-off lands, viz., China, Korea, Burma (now Myanmar), Sinhala (now Sri Lanka), Nepal, and Tibet. In the current date, India manages one of the largest higher education systems in the world (UGC, 2017).

Higher education in India has been the subject of criticism of many for several reasons, viz., poor quality course content, shortage of qualified teaching faculty, lack of research interest, inadequate infrastructure facilities, limited financial support, uneven industry-oriented skills, poor international collaborations, lack of motivation to compete internationally, meager research output and number of citations, reluctance to establish global Universities, and so on (Reddy et al., 2016). The engineering education in India is also not free from such allegations, which also explain some of the limitations regarding India not able to jump to the next level in the arena of manufacturing engineering. Due to the lack of advanced manufacturing techniques, India still has to depend upon other countries in this aspect. Precision manufacturing is another sector where India is lagging behind many developed nations of the world. Although there is no magic potion to change this scenario with a wink of an eye, improvement in this sector heavily depends upon the quality of the manufacturing workforce India is getting. A quality workforce can only be produced by a revamped manufacturing education scenario in the country. After all, improved education system would give rise to strengthened manufacturing sector which will expand the country's economy and propel it toward the path of progress.

2 ABOUT MANUFACTURING ENGINEERING

Manufacturing engineering is one of the engineering disciplines which deals with various manufacturing processes. Manufacturing involves turning raw materials into finished products to be used for various purposes. It is an association of mechanical engineering, industrial engineering, electrical engineering, electronic engineering, computer science, materials management, and operations management. This stream was introduced in the mid to late 20th century when industrialization was at its peak. As industrialization gathered pace, requirement was felt for specialized people who can take over the manufacturing department of the industries. From this aspect, a separate discipline of Manufacturing Engineering was introduced and, within a short span of time, it became popular with the students.

Manufacturing Engineering is a branch of professional engineering concerned with the understanding and application of engineering procedures in manufacturing processes and production methods. Frequently, Manufacturing Engineering is considered as a subset or specialization of Mechanical Engineering with the focus only on machine tools, materials science, tribology, and quality control. However, some institutes offer Manufacturing Engineering or Production Engineering as an independent engineering discipline.

Manufacturing is the most important element in any engineering process and Manufacturing Engineers are key personnel in many organizations. The manufactured products range from aeroplanes, turbines, engines, and pumps—to integrated circuits and robotic equipment. Professional manufacturing engineers are responsible for all aspect of the design, development, implementation, operation, and management of manufacturing system. Under Manufacturing Engineering discipline, various courses are taught through engineering colleges and Universities. Degrees like Bachelor in Engineering (B.E.), Master in Engineering (M.E.), and Research degrees like Ph.D (Doctor of Philosophy) are awarded. Both in India and abroad, a degree in Manufacturing Engineering is considered both an academic qualification and technical qualification (Indian Education, 2017).

The services of a manufacturing engineer are sought in all industries producing goods—from automobiles, ships, airplanes, to electronic goods and educational toys, to food and clothing. A manufacturing engineer must have eyes for details. He/she needs to be a team worker; both with engineers and technicians two- or three-tier down the supply chain. A Bachelor of Engineering (B.E.) degree is the minimum eligibility criteria for entry-level positions as a manufacturing engineer.

2.1 Responsibilities of Manufacturing Engineer

Manufacturing or Production Engineer's primary focus is to turn raw material into an updated product or a new product in the most effective, efficient, and economic way possible. For this, he/she works toward choosing machinery and equipment for the particular manufacturing process. The actual responsibility of such an engineer is listed below.

- Manufacturing engineers will be planning and scheduling the production in any manufacturing industry (e.g., Automobile Manufacturing industry).
- Manufacturing engineers will be programming the CNC machines to produce engineering components such as gears, screws, bolts, etc.
- They are responsible for quality control, distribution, and inventory control.

Manufacturing engineers also work with the tools like programmable and numerical controllers, robots, and vision system to fine-tune assembly, packaging, and shipping facilities. In general, manufacturing engineer works with a model or, say, prototype created with computers to execute the final manufacturing process. Manufacturing engineer comes out with methodologies to manufacture a product in efficient and cost-effective manner, and hence, create a marketing edge for the final product.

2.2 Courses and Eligibility

Manufacturing Engineering as undergraduate and postgraduate program is offered in various Universities and colleges. Minimum eligibility criterion for studying Manufacturing Engineering, as undergraduate program, is a higher secondary $(10+2)$ or equivalent qualification with physics, chemistry, mathematics, and english. For admission to graduate programs, all India joint entrance exams are conducted. Some states conduct their own entrance tests for admission. For admission into postgraduate programs, qualification in GATE (Graduate Aptitude Test in Engineering) is in general the eligibility criteria. However, various colleges and Universities have their own set of rules for admission.

2.3 Career Prospects

Manufacturing engineering graduates can find jobs in all those industries where something is manufactured, processed, packaged, or shipped. Both government and public sector enterprises involved in manufacturing of goods and commodities need services and expertise of manufacturing engineers. These industries can range from needle makers to ship builders, to pharmaceuticals or electronic chip producers. Manufacturing engineering graduates can find job opportunities in educational settings and R &

D (Research & Development) laboratories. Work opportunities in industry may include the following work areas:

• Process/product development
• Process improvement
• Manufacturing process engineering
• Mechanical process design
• Mechanical analysis
• Utilities, power generation, and power distribution
• Plant support engineering
• Research
• Technical sales and service
• Information Technology

When one has a B.E. in manufacturing engineering, he/she can serve the industry as manufacturing engineer or can join a master's degree program in India or abroad or can appear for competitive exams like civil services (Indian Education, 2017). So, the opportunities are enormous for a manufacturing engineering graduate. Moreover, manufacturing engineers being the soul of the manufacturing backbone of any country can serve in its progress and development. To sum up, one can say manufacturing engineering is an evergreen field.

3 GROWTH AND EDIFICE OF ENGINEERING EDUCATION IN INDIA

From the establishment of the first engineering college named Thomson Engineering College in Roorkee, India has seen a sea change in engineering education all over the country. At the end of 19th century, India had only four degree level engineering colleges, about 20 survey and technical institutes, and about 50 industrial schools (Choudhury, 2016). In 1945, Government of India established higher technical institutes based on the lines of Massachusetts Institute of Technology in different regions of India, which came to be known as Indian Institute of Technologies (IITs) at Kharagpur (1950), Bombay (1958), Kanpur (1959), Madras (1960), and Delhi (1961). These IITs, together with Indian Institute of Science (IISc), are considered the premiere hubs for engineering education in the country today. Moreover, due to increased demand for technical education, the Government had set up several Regional Engineering Colleges (now known as National Institute of Technologies (NITs)) across the country. Apart from these, several States have set up their own engineering institutions which have gained popularity over time. Jadavpur University, Anna University, etc., are some of the State-funded institutions. Several private organizations have also set up engineering colleges which have delivered quality education over time.

There has been a phenomenal rise in the state of higher education in India post-independence. From 500 colleges and 20 universities at the time of independence, the number of colleges and universities has increased by a factor of 82.87 and 42.35, respectively. Consequently, the gross enrolment rate of students in the formal system of higher education has gone up by 135.64 times (UGC, 2016). Among these enrollments (higher education), about 10% belong to the professional courses, viz., engineering and technology. This implies that India currently is producing one of the largest shares of engineers across the globe. Fig. 1 shows the faculty-wise distribution of universities in India in year 2015–16. Table 1 shows faculty-wise student enrollment in India during the same year.

India has witnessed a boom in engineering education in the last two decades. During 1991–2001, the rate of increase in the number of institutions of higher education (including all disciplines) was around 8%, whereas the same in case of engineering institutes was about 15% (Choudhury, 2016). This also led to sevenfold jump in the number of students enrolling for engineering colleges compared to only threefold overall enrollment increase during the same period. The total number of enrollments for engineering courses is about 49 lacs (Table 1). This means a large amount of engineers are produced each year in India. It is further noticed from surveys

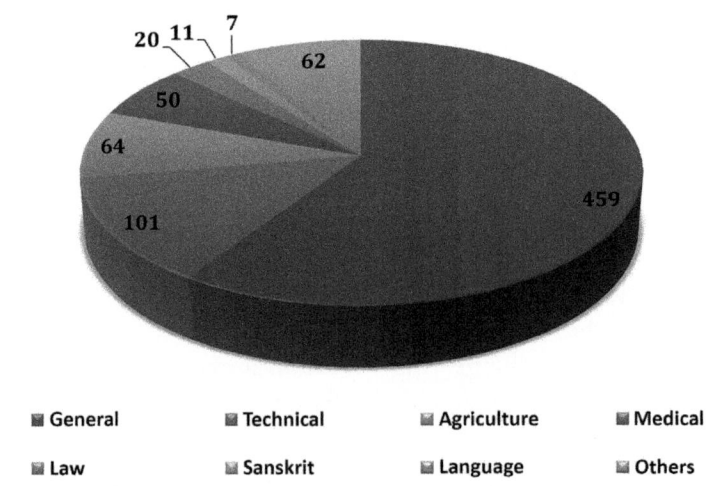

Fig. 1 Distribution of Universities as per specialization (GoI, 2016).

Table 1 Faculty-wise student's enrollment 2015–16 (UGC, 2016)

S. no	Faculty	Total enrolment	Percentage to total
1	Science	5,417,464	19.02
2	Arts	10,271,296	36.06
3	Commerce/Management	4,637,317	16.28
4	Engineering/Technology	4,885,134	17.15
5	Medicine	1,118,178	3.93
6	Agriculture	240,090	0.84
7	Education	1,085,876	3.81
8	Veterinary Science	31,332	0.11
9	Law	474,423	1.67
10	Others	323,636	1.14
Total		28,484,746	100.00

(GoI, 2016) that more than 70% students in engineering are males and female participation is relatively low. All the technical institutes together offer undergraduate (UG) engineering degrees in 45 major disciplines. The top five popular engineering streams on the basis of enrollment are Mechanical Engineering, Computer Engineering, Electronics Engineering, Civil Engineering, and Electrical Engineering (GoI, 2016).

Because of the large number of engineers produced, India faces a challenge in maintaining the quality of education at such a large scale. Besides, a number of issues, viz., regional disparities in the growth of institutions, public financing of engineering and technical education, and uneven access by gender and social groups, have come up in a serious way and thus demand for a suitable planning and management (Choudhury, 2016). In this context, it should be mentioned that AICTE (All India Council of Technical Education) is a statutory body under Government of India which is responsible for proper planning and coordinated development of the technical education and management education system in India. The vision of AICTE as mentioned in their official webpage (AICTE, 2017a) is "To be a world-class organization leading technological and socio-economic development of the country by enhancing the global competitiveness of technical manpower and by ensuring high quality technical education to all sections of the society." The objectives of the body are:

- Promotion of quality in technical education
- Planning and coordinated development of technical education system
- Regulations and maintenance of norms and standards

In addition to AICTE, UGC (University Grants Commission) is charged with coordination, determination, and maintenance of standards of higher education. It provides recognition to Universities in India and disburses funds to such recognized universities and colleges. Ministry of Human Resource Development, Government of India keeps an overall watch upon the education system in India.

In India, technical education is primarily disseminated through two courses, viz., undergraduate (UG) degree course and diploma (polytechnic) course. However, for advanced technical knowhow and research, postgraduate (PG) courses and doctoral degree are also becoming more and more relevant and popular. Apart from these, for getting training in different industrial trades, ITI (Industrial Training Institute) courses are also taken up by many students who are unable to make it to UG or polytechnic courses. Among these, Polytechnic and ITI courses can be pursued at postsecondary level, while UG degree course can be attended only after higher secondary (Class 12). Joint Entrance Examination (JEE) is conducted by the Central Government (by CBSE, GoI) for admission to Central-funded institutes, viz., IITs, NITs, etc. State Governments also conduct their own JEE for admission to state-funded colleges and Universities. Private technical institutions, in general, follow the merit list of both the entrance examinations for admitting students at UG level. However, some conduct their own entrance tests. Admission at PG level is preferably done based on the scores obtained in GATE (Graduate Aptitude Test in Engineering) examination (conducted by one of the IITs in rotary manner). However, some institutes admit students through their own tests and interview. Doctoral admission is normally done based on the academic background of the student together with an admission test (based on the field selected by the candidate) and interview by the institute.

The higher education in India is mainly based on the Semester system. Duration of diploma course is six semesters (3 years), while that of UG engineering (B.Tech. or B.E.) course is eight semesters (4 years). PG course (M. Tech. or M.E.) is four semesters (2 years) long with the last two semesters dedicated fully for project work. Duration of Doctoral program normally ranges from 3 to 5 years, but can be extended according to need of the research work which is regularly monitored by the Supervisor. End semester exams are conducted for evaluation purposes in case of diploma and UG courses, whereas end semester exams together with the project work is employed for evaluating a student at PG level. In case of doctoral students, a successful thesis defense in the presence of an external expert together with research publications in journals and conferences provides a means of evaluating the candidate.

4 GLOBAL POSITION OF INDIA IN SCIENCE AND ENGINEERING EDUCATION

In the present world which is very much interconnected, workforce with science and engineering (S&E) skills are integral for innovation and economic competitiveness. In many countries including India, Governments have made increased access to S&E-related postsecondary education and made it a priority (National Science Board, 2016). It is interesting to note that these types of degrees have become relatively more prevalent in some Asian countries than in the United States. Globally, the number of first University degrees reached about 6.4 million in 2012. As seen from Fig. 2, India (23%) and China (23%) together conferred almost half of these degrees. However, mere higher number of S&E degrees does not indicate the knowledge-driven development as it can also be related to large population in these two countries. Besides, understanding the relationship between degrees conferred in a country and the capabilities of its workforce is complicated as increasing numbers of students are receiving higher education in foreign countries. Among many countries, the United States remains the preferred destination for higher education of a large number of students in India.

In case of research, the level of doctoral enrollments in India is quite low compared to that with the developed nations of the world. Engineering and Technology remained third in terms of enrollments after arts and science

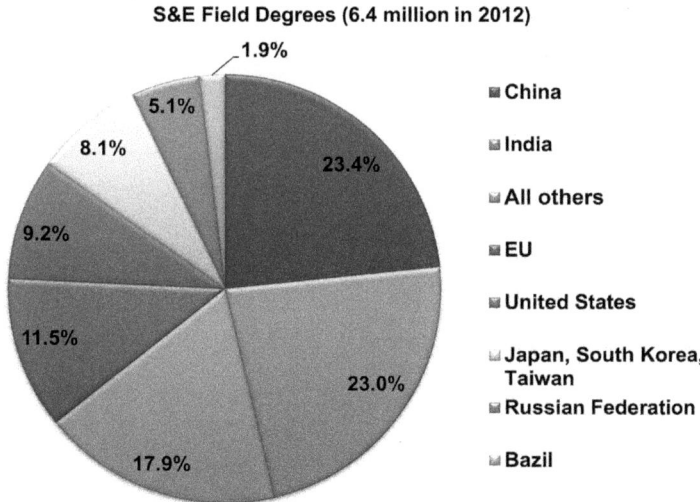

Fig. 2 First University degrees, by selected region/country/economy: 2012 (National Science Board, 2016).

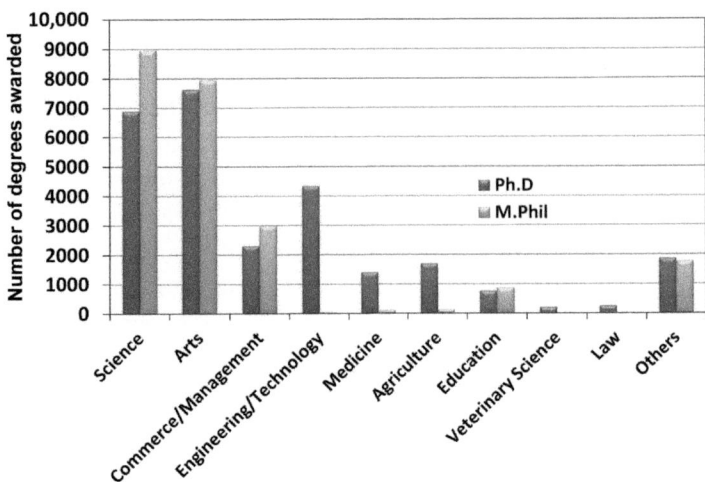

Fig. 3 Faculty-wise number of M.Phil and Ph.D students in India during 2014–15 (UGC 2016).

(Fig. 3). It is further found that in case of Engineering and Technology courses, 11.7% students opted for Ph.D after their postgraduation. It is also observed that Mechanical Engineering remains the most popular choice among the students even at the doctoral level.

There is a need to give thrust on the manufacturing engineering education in various entities, including especially institutions of higher education or consortia of such to strengthen their technical programs so that students could meet the demands of modern manufacturing industry. Manufacturing leaders, Germany, Japan, and China, are moving toward promoting manufacturing engineering education. Hence, India must take timely steps to move things forward.

5 IMPORTANCE OF MANUFACTURING ENGINEERING EDUCATION

5.1 Effect of Manufacturing on GDP

GDP (Gross Domestic Product) is a very strong measure to gauge the economic health of a country and it reflects the sumtotal of the production of a country and, as such, comprises all purchases of goods and services produced by a country and services used by individuals, firms, foreigners, and the governing bodies (Jain et al., 2015). It serves as an indicator for all the governments and economic decision makers for planning and policy formulation.

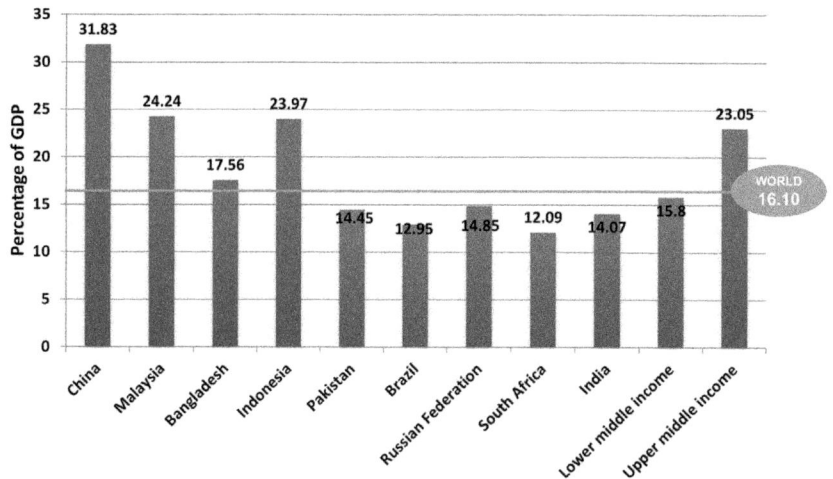

Fig. 4 Comparison between countries based on manufacturing as a percentage of GDP in 2012. *(Source: World Bank, CMIE.)*

For any future planning concerning the nation, various sectors' contribution to the GDP is determined.

The GDP of a country is dependent on several sectors, manufacturing sector being one of them. Developed countries display a heavy contribution of the manufacturing sector in their GDPs. Indian GDP is mainly dominated by contributions from agriculture and services sector, which together constitute around 75%. The manufacturing sector's growth stood at 9% of India's GDP in 1950–51, stagnated around 15% for over two decades (Revathy, 2017). But these figures are still low compared to India's neighboring countries (Fig. 4), viz., China (31.83%), Malaysia (24.24%), and Indonesia (23.97%). The Indian manufacturing share (14.07%) is even lower than the world average (16.10%). Hence, a boost in manufacturing sector is required so that India can catch up with its neighboring countries. This boost is dependent on the availability of a properly trained and efficient workforce which is possible only through a good manufacturing engineering education system.

5.2 Defense Applications

India maintains a strong and heavy defense force from security point of view. There is a regular demand of defense-related equipment, viz., battle tanks, missiles, fighter planes, drones, etc. However, due to an underdeveloped defense manufacturing sector, India is one of the largest importers of conventional defense equipment in the world. India imports approximately 60% of its defense requirements. This makes India's defense sector one of the most attractive markets globally for both domestic and foreign defense manufacturers.

Defense is the best prospect industry sector for this country, provided India has the necessary infrastructure and workforce to tap it. Realizing this, Government of India has made it a priority to create a robust defense industrial base under its "Make in India" initiative, which also aims to create jobs in India and further develop its overall manufacturing capabilities. Hence, to grab this opportunity, India must also produce competent manufacturing engineers and technicians, which can be possible only by a robust manufacturing engineering education system.

5.3 Space-Related Applications

The Indian space industry is grabbing eyeball, thanks to its premier space organization ISRO (Indian Space Research Organization). India is one among seven countries in the world which can launch a full-fledged satellite from its own soil using its own launch vehicle. Riding high on the successful Moon and Mars mission, India has seen a flurry of demands from foreign countries to launch their satellites. The chief reason for this is the reliability of the Indian launch vehicles and the low cost of launching. Even many developed countries look up to India for putting their satellites in orbit. Hence, ISRO has had to increase the number of launches to meet the demand. Thus, space-based industry is another lucrative sector which India is trying to capture. There has been growth of a number of industries who manufacture various rocket parts which are then assembled by ISRO. Of late, there is a plan to hand over some of the proven launch vehicles to the industry, so that the Organization can devote more attention towards Human Missions and Interplanetary Missions. Thus, there exists a need of efficient manufacturing workforce who can manufacture these sophisticated parts for space-related applications.

6 STATE OF MANUFACTURING ENGINEERING COURSES IN INDIA

The basis of any industry, be it related to metal works, manufacturing of electronics product or computer hardware, etc., is manufacturing process. Manufacturing science or manufacturing technology is offered in the first year undergraduate studies (all engineering disciplines) as a subject which includes the metal and it's alloy properties, metal forming, casting process, welding process, and different applications of manufacturing engineering in the industry (Khare et al., 2016). It is thought to be quite relevant to include a case study in order to unravel more clearly the actual scenario prevalent. Jadavpur University is thought to be an ideal candidate as it maintains a descent position among the technological institutions across the country (NIRF, 2017).

6.1 Case Study—Jadavpur University

The inception of Jadavpur University can be traced back to 1905–06 before the independence of India. After independence, the Government of West Bengal, with the concurrence of the Government of India, enacted the necessary legislation to establish Jadavpur University in 1955. At present, Jadavpur University has successfully established itself as a foremost Indian University with a vast repertoire of courses offered, an enviable list of faculty members, and has come to be known for its commitment toward advanced study and research. Hence, Jadavpur University can act as a broad representative of a majority of the engineering colleges in India where a certain standard of education is maintained. Being a state University, it does not enjoy the financial generosity like the central government institutions, viz., IISc, IITs, and NITs. Moreover, it is challenged by various regional, social, and political issues time to time. Nevertheless, the University has produced many competent engineers who have become leaders in their fields of interests.

In Jadavpur University, almost all the popular engineering disciplines are taught, viz., Chemical Engineering, Civil Engineering, Computer Science & Engineering, Construction Engineering, Electrical Engineering, Electronics & Telecommunication Engineering, Information Technology, Instrumentation & Electronics Engineering, Mechanical Engineering, Metallurgical & Material Engineering, Pharmaceutical Technology, Power Engineering, Printing Engineering, and Production Engineering. Apart from these, engineering streams, viz., Food Technology & Bio-Chemical Engineering as well as Architecture, are also offered. It is interesting to note the presence of the Production Engineering stream, which is totally dedicated to manufacturing engineering. However, Mechanical Engineering course also contains a lot of manufacturing-related subjects. All the disciplines offer degrees from UG level to doctoral level.

6.1.1 Manufacturing Engineering Education at UG Level

In Jadavpur University, manufacturing engineering education at the UG level is mainly categorized into practical as well as theoretical subjects. Practical subjects are mainly related to train the students with the general workshop practices. Jadavpur University has a very good workshop facility with separate sections for carpentry shop, foundry and forging shop, welding shop, and machine shop. The machine shop popularly referred to as the "Blue Earth Workshop" is one of the oldest and largest of its kind in the country.

In the first year of almost all the disciplines, basic workshops courses like carpentry, molding & casting, forging, and machine tool operations are conducted. These courses are limited mainly up to second year, except for Mechanical and Production Engineering streams which offer detailed and extensive workshop practices going up to fourth year of their respective programs. Moreover, the theoretical subjects related to manufacturing engineering are also mainly limited to these two disciplines. However, as already discussed, manufacturing requires knowledge of almost all the domains. Hence, the core principles that are taught in other branches of engineering, viz., Electronics, Electrical, Chemical, etc., are equally important to manufacture interdisciplinary products, viz., electronics hardware, computer hardware, plastic products, etc. The present discussion about manufacturing engineering education is limited to Mechanical Engineering stream as the same is very popular and widely offered in various engineering institutions all over the country.

As already mentioned, the manufacturing engineering-related subjects in Mechanical Engineering are broadly divided into practical workshop-related laboratories as well as theoretical subjects. There are five workshops offered over the entire duration of eight semesters of the UG program. These workshop starts with basics of carpentry and pattern-making where students get an experience in handling carpenter's tool with which they make a primitive job with wood. Next, the students undergo introduction to Fitting and Welding processes. Students get hand-on experience in various welding techniques, viz., electric welding, gas welding, tungsten inert gas (TIG) welding, metal inert gas (MIG) welding, etc. Moreover, fabrication techniques, viz., soldering and brazing, are also taught to the students. In forging and molding shop, students are required to do basic forging operations as well as prepare molds for casting. Finally, the students are exposed to basic and advanced operations of various industrial machine tools, viz., lathes, drilling machines, shaping machine, slotting machine, milling machine, and grinding machine. Students get hands-on experience in operation of lathe where they make simple metallic jobs. In the advanced machine laboratory, students are required to manufacture the components of a complete equipment or device like centrifugal pump. They carry out fitting, machining, assembly of the manufactured components, and finally carry out testing of the device manufactured. Besides, manufacturing of special components, viz., screw, gears, etc., are also undertaken. Experiments in metal cutting, viz., study of chip formation mechanism and influence of various factors, tool life, etc., are also done. Students are also introduced to modern machines like CNC (Computer Numerical Controlled) lathe and nonconventional machining methods, viz., Electro-discharge Machining (EDM). Besides special laboratories on metallography and metrology are offered where students prepare samples for heat treatment and undertake metallographic testing of metals and study their microstructures. In the metrology part, students learn about various measurement techniques prevalent in the industry. They also learn to handle equipment, viz., surface profilometer, microscope, use of slip gauges, etc.

In case of theoretical subjects, topics, viz., Machining Technology and Metrology, Materials and Metallography, Machine Tools, and Advanced Manufacturing, are included in the UG Mechanical Engineering curriculum. These topics together cover almost all the basics of manufacturing engineering starting from selection of materials to advanced manufacturing topics like computer-aided design and computer-aided manufacturing. Even a topic called Mechanical Handling of Materials is also included which gives the students an idea about how objects, big or small, are suitably handled in the industry so that work is carried out smoothly within the least amount of time. There is also a rich pool of elective subjects (related to manufacturing engineering) available to the students, which help in supplementing their primary knowledge as well as providing advanced information about a particular area of manufacturing engineering; topics include Mechanical Measurements and Industrial Statistics, Introduction to Industrial Pollution, Computer-Aided Design and Manufacturing, Advanced Production Processes, Laser Machining Process, Production Systems and Controls, etc. Students are also taught about the basics of production management techniques through subjects, viz., Engineering Economics and Costing, Industrial Management, Operations Research, etc. There is also a compulsory industrial tour for the students in which they have to visit a few industries nearby and submit a report and give a presentation for which they are given credits. The students are made to go through a proper evaluation system, which includes a Semester examination (at the end of each semester) as well as a series of tests and viva voce examinations.

6.1.2 Manufacturing Engineering Education at PG and Doctoral Levels

At PG level, students are offered courses which address the manufacturing processes in a detailed manner. Topics like Advanced Manufacturing Science and Advanced Manufacturing Processes are taught to the students. Students are also exposed to the concepts of Manufacturing Aspect of Design. It requires the students to apply the basic knowledge of manufacturing processes in order to properly design a product which would be suitable and also economical to manufacture. This subject teaches the student to assess the feasibility of manufacturing from his designs. The student also develops the ability to choose the appropriate manufacturing process among various available choices so that a part can be manufactured in the most economical way and without compromising on its quality. Moreover, the student may also modify the design marginally to suit a particular manufacturing process which may be readily available. Some research works related to manufacturing science are also undertaken by the students at the Doctoral levels. Both experimental works involving concepts like design of experiments in order to optimize some parameters to obtain the desired response and management-related topics are undertaken. Despite limitations in funding and unavailability of cutting-edge equipment, publications in International Journals and participations in Conferences are regularly done by the students. The library of the University is well-stocked with a range of books related to manufacturing engineering. Besides, the University also has a variety of subscriptions of relevant e-contents, journals and magazines, etc., which are easily accessible to the students.

6.1.3 Use of Simulation Tools

Almost all the major simulation tools required in mechanical design are available in the University to which the students have easy access. Programming languages, viz., FORTRAN, MATLAB, SCILAB, etc., have been introduced in UG curriculum. By learning these tools, students are able to undertake simple mathematical problems related to engineering and solve them. Besides, computer-aided drafting, which is now an essential part of any industry, is also an integral part of the engineering curriculum. Drawing using AutoCAD is gradually replacing the manual drawing, especially for the advanced machine drawing classes. Students are also exposed to the solid modeling software, viz., Creo Parametric, Solidworks, etc., with which they are able to model simple mechanical assemblies and check for any interference. Keeping with the demand of time, Finite Element Analysis using tools, viz., ANSYS, ABAQUS, etc., have been introduced in the curriculum.

However, all these tools are extensively used by the PG and Doctoral students in their research work. With limited facility for experimentation, lack of sophisticated equipment, and other budgetary issues, these simulation tools have come as a relief to the students carrying out research works. Time to time, training programs on these softwares are arranged by the Department.

Hence, looking at the various manufacturing-related topics taught under the umbrella of mechanical engineering, it can be inferred that there is almost no lacunae as per as the syllabus and course curriculum is concerned. Almost all the topics which a student (especially UG level) is expected to know about manufacturing engineering is covered in sufficient details. However, the course content alone is not responsible for an efficient teaching-learning process. For this, the quality of students and teachers is equally important. In case of Jadavpur University, recruitment of teachers is done in a rigorous manner satisfying all the regulations of UGC, the Government of India authority which looks after the quality of higher education in India. Almost 100% of the faculty members in the University are doctorates with a very good academic background. Moreover, being a premier state University, Jadavpur University attracts the top ranked students from the Joint Entrance Examination. Hence, the overall scenario related to manufacturing engineering education at the University looks favorable for inculcating the relevant knowledge into the students so that they can be competent engineers in today's industry.

7 LACUNAE IN THE MANUFACTURING ENGINEERING EDUCATION SYSTEM

Presently, India operates as one of the largest engineering education systems in the world and churns out a large number of engineers annually. However, it is a worrying fact that most of the graduates are not employable. According to survey conducted by McKinsey & Company (NASSCOM-McKinsey, 2011), some time back only about quarter of the engineering graduates in India are employable for technology-related services. This general employability condition of engineering graduates has direct impact on the country's manufacturing sector. Some of the reasons for this poor employability are outdated curriculum, poor teaching infrastructure, and shortage of good faculty, particularly in institutes lower down the order. Following text attempts a detailed discussion about the reason for this dismal performance of the engineering graduates.

7.1 Poor Infrastructure in Private Technical Colleges

The onus of bringing out quality engineers lies not on the students, most of whom come into the system with the intention to learn, but on the institute. In India, majority of the engineering graduates come from the private colleges. Hence, these colleges control the quality of the technical workforce who are also responsible for the progress of the manufacturing sector and the overall development of the country. Now, there has been a significant change in the character of private partners in higher and technical education in India. Post-independence in around 1950–60, kind-hearted affluent people used to donate their money to public institutions or provided help in setting up of philanthropy-based private schools and colleges (Choudhury, 2016). However, today in the money-minded society, people with even a small fraction of money in hand prefer to set up a private, self-financing college or University. The investment in higher educational institutes is found to be the most rewarding, yielding quick and very high pay-offs, with minimum risk (Choudhury, 2016). This has led to the mushrooming of especially engineering and management colleges, which with some notable exceptions have largely become mere business entities dispensing very poor quality education (Yashpal Committee Report, 2008).

The philanthropy and charity motives of investing in higher education have been replaced with profit making and commercial interests (Choudhury, 2016). There has been remarkable establishment of profit-oriented commercial institutions in India, especially after 1990. However, the real explosion of these types of technical institutions happened after 2000, thanks to the liberal regulations and policies of the Government. Many of these institutions, although having publicized philanthropic visions and missions declaring themselves as not-for-profit institutions, contributed to vulgar forms of commercialization in technical education in India. Riding on the boom in Information Technology in India, there was a sudden requirement of software and hardware engineers. Many private entities during that time opened colleges with some computational facility and offered engineering streams such as Information Technology, Computer Science and Engineering, and Electronics and Communication Engineering. There was no mechanical workshop and the basics of manufacturing were entirely taught theoretically with the students having little or no practical exposure to the basic machinery. With time, to make the establishment more profitable, core engineering branches, viz., Mechanical Engineering, Electrical Engineering, etc., were opened with limited laboratory facilities. Some colleges tied up with Government technical institutions and took their student to a visit to the workshops of those institutions. It has also been reported that majority of privately funded and managed engineering institutions are engaged in malpractices such as collecting exorbitant capitation fees and other institutional fees and manipulation of entrance results and admission processes, such as disregarding admission norms in favor of those willing to pay more (Choudhury, 2016). Moreover, luring students with placements which are many times fixed have also been reported. After these dubious placements, mostly the candidate is made to sit in the bench or kicked out of the company after a period.

7.2 Disparity Between Tier-1, Tier-2, and Tier-3 Engineering Colleges

In continuation with the poor infrastructure of the private technical institutes, it is thought relevant to discuss the various points of disparity between different tiers of colleges in Indian engineering education landscape. For academic purpose, these colleges are classified into Tier-1, Tier-2, and Tier-3 categories. If a college complies with the Washington accord, it is more likely graded as a Tier-1 institution. For a Tier-2 college, an "A" certification from National Assessment and Accreditation Council (NAAC) is a requisite. Engineering Colleges failing to meet either of the aforementioned criteria constitute the list of Tier-3 engineering institutes.

By default, IITs, NITs, and other Central Government-funded colleges come under the Tier-1 category, while the Tier-2 lists accommodate several private and state-funded institutes. Apparently, Tier-1 colleges vis-a-vis Tier-2 and Tier-3 ones are elite members in the arrangements. They receive generous grants from the union government, which accounts for their thriving infrastructure, better faculty, and cutting-edge teaching techniques. In addition, recruiters prefer these institutes to diversify their talent pool. Unfortunately, their poorer cousins, the Tier-2 and Tier-3 colleges, are devoid of any such privileges.

The majority of Tier-2 and Tier-3 campuses paint a dismal picture. They are devoid of even the basic necessities like clean drinking water, sanitation, lecture halls, and labs. Ill-equipped labs and classrooms, low grade IT setup, and small sized hostel accommodations are hallmarks of colleges with a "decent" infrastructure. The grants these institutes receive are either underutilized or inappropriately utilized or is insignificant enough to bring about a noticeable change.

The disparity between the colleges is heavily reflected in their campus placements. While the Tier-1 counterparts grab top jobs and healthy remunerations and benefits, the Tier-2 and Tier-3 college alumni may not find much favor with recruiters. The possible reasons for such apathy are the inferior quality of education, lack of exposure, negligible focus on faculty-student interaction, and insufficient infrastructure which has already been deliberated. Additionally, a significant percentage of Tier-2 and Tier-3 alumni lack communication abilities and their verbal and written English skills are deplorable. Not many colleges make efforts to bring about a change, and hence, students are left to fend for themselves. Moreover, the absence of placement cells in most institutions also restricts students' placement goals.

7.3 Outdated Teaching Methodology and Curriculum

At the undergraduate level, the purpose of an engineering curriculum is not to make one current in any particular field. That is a constantly moving goal-post. Rather, the purpose of an undergraduate engineering curriculum is to develop a toolbox and skills that can then be used to solve problems in any particular field. Mostly, basic math and science do not change: calculus, stoichiometry, statics and dynamics, stresses and strains, strength of materials, thermodynamics, etc., do not really change. Hence, many times the subject matter need not to be changed, but the implementation needs to be changed in order to provide more motivating context. For example, engineering curriculum needs to have courses in programming to enable the students develop algorithm skills. A majority of the colleges continue with the age-old curriculum which is many times deemed irrelevant by the industry where the students are supposed to get placed. Now, apart from the conventional manufacturing techniques, the field of manufacturing is advancing everyday which is reflected in the availability of the sophisticated equipments that we have incorporated in our day-to-day lives. However, many Universities are not proactive to put into place a regular syllabus revision mechanism so that more contemporary topics related to manufacturing engineering could be included. Moreover, topics, viz., automation, cloud, data analytics, internet of things, etc., should be incorporated in the curriculum keeping with the demand of time.

7.4 Unavailability of Qualified Faculty

A good teacher enlightens the ignorant mind with fire of knowledge and also contributes to the development and prosperity of a nation. However, a person cannot be a good teacher unless he has acquired the knowledge himself and has the necessary communication skills to convey the same to the students. While UGC has mandated M.Tech. or M.E. to be the minimum qualification of the teacher of an engineering college, many of the colleges are managing teaching with B.Tech degree holders who are many times pass-outs of the same college. Even nondoctoral candidates are promoted to sit in the post of the Principal violating the rules and regulations. A majority of the faculty members also do not undergo orientation and pedagogical courses which are essential for development of one's teaching abilities.

Except a few colleges, the number of faculty members with Doctorate degree is numbered and those too many times are retired and old persons unfit for teaching. According to the report published by the National Institute of Ranking Framework (NIRF, 2017), a little more than a quarter fraction of the engineering faculty engaged in teaching has doctoral qualifications (Fig. 5). While it is true that in a few disciplines that may not be a serious handicap, in many cases the mentorship received during the doctoral programs plays a key role in preparing the faculty for a teaching career in higher education, and the diffusion of this trend needs clearly to be speeded up.

In an engineering institute, especially when mechanical or manufacturing engineering education is concerned, experience of working in a relevant industry is believed to be helpful for the faculty. This industrial exposure enables the teacher to oversee the link between theoretical topics and their practical applications which he can then bring up in the class creating more interest and enthusiasm among the students. It is unfortunate that such type of faculty members is very limited in numbers in the engineering education system of the country. Another major trend that is clearly visible (Fig. 6) is

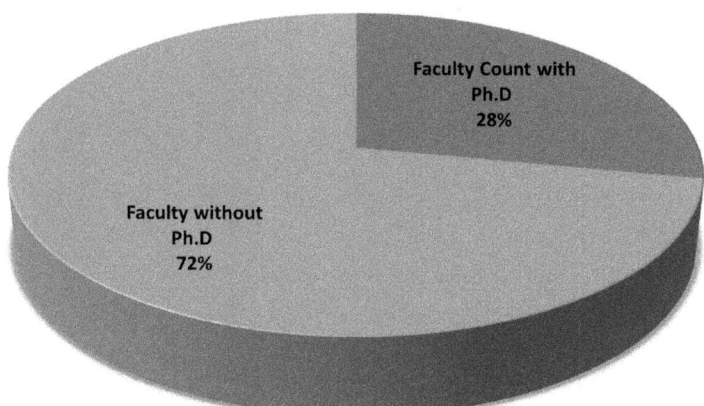

Fig. 5 Comparison: Engineering faculty with and without doctoral degree (NIRF, 2017).

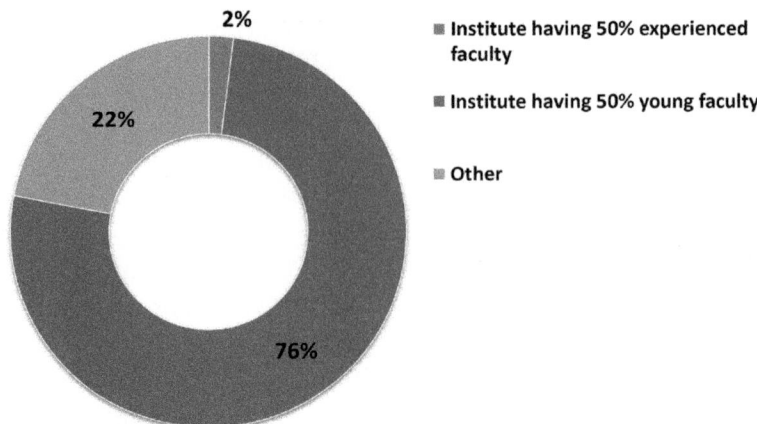

Fig. 6 Faculty experience in engineering institutes (NIRF, 2017).

that majority of institutes (nearly 76% of them) have significant number of faculty with less than 10 years of teaching experience (about almost 50%)— in other words, the burden of the teaching is largely in the hands of relatively inexperienced faculty, which puts another serious question mark on its impact on the quality of education (NIRF, 2017).

Another factor which limits the technical education is the faculty-student ratio (FSR). In an overpopulated country like India, maintaining the ideal FSR ratio is difficult. NIRF survey shows that across the Engineering categories, even among the participant institutes, which represent among the most aspirational 3000+ in the country, a significant number have a long way to go to have reasonable numbers here. Some are working with less than a teacher for 50 students or more (Fig. 7). In a field like manufacturing engineering, this greatly limits the quality, since it implies that a faculty member is highly loaded and expected to teach heavily across different kinds of courses (NIRF, 2017).

7.5 Lack of Industry-Oriented Research and Innovation in S&E in India

Research- and development-related works are very necessary for a country in order to keep pace with the rest of world in the current scenario when technologies are improved every day. Without the implementation of cutting-edge technologies, the economic development of a nation would definitely suffer. It is unfortunate that a country like India is still lagging behind in this aspect as can be observed from the regional expenditure in

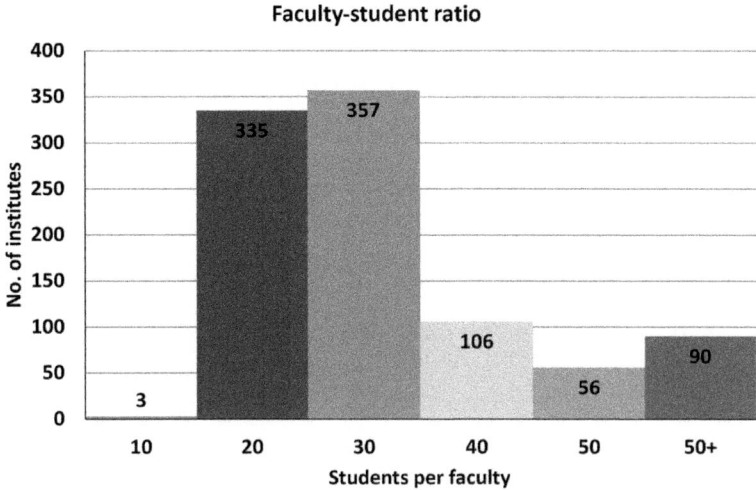

Fig. 7 Faculty-student ratios in engineering institutes (NIRF, 2017).

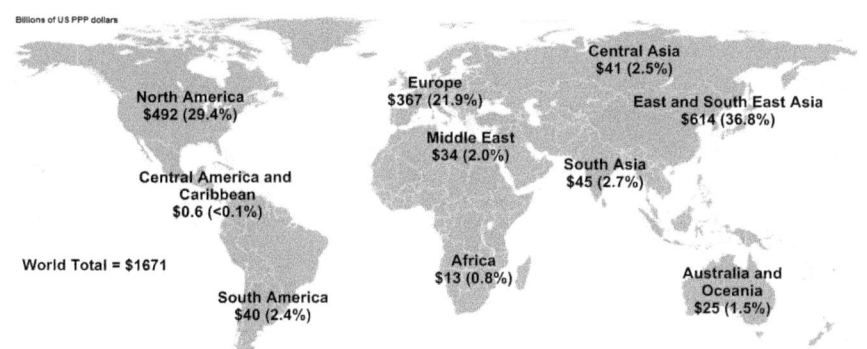

Fig. 8 Global R&D expenditure (National Science Board, 2016).

R&D across the world (Fig. 8). The lower share of South Asian countries is an indication of relatively lesser number of researchers as well as labor force who are involved with this type of works. Research works are concentrated only at the top technological institutes and some of the government-funded R&D organizations and laboratories. This also implies very less people having potential of joining manufacturing workforce are acquainted with the cutting-edge technologies. A major share of the workforce continue with the traditional methods of manufacturing even if they have been deemed obsolete in advanced countries, viz., America or Japan. This definitely imposes a drag in the overall development of the country.

Some of the organizations involved in cutting-edge research in India are ISRO, Defense Research and Development Organization (DRDO), and Department of Atomic Energy and various laboratories, viz., Council of Scientific and Industrial Research laboratories, which carry out R&D-related works regularly. Apart from these, a few private organizations have their own laboratories. In case of academia, mainly IITs and IISc with a few other institutes are recognized for some degree of R&D environment. But majority of the remaining technical institutes, be it of graduation or diploma level who actually churn out a huge number engineers and technicians who would become part of the manufacturing workforce, are distances away from any type of R&D. This is to be blamed somewhat upon the lack of awareness and initiative of the respective authorities. Moreover, there is no mechanism for communication and sharing of technological knowhow among the institutes and the research centers.

Refereed S&E journals provide a tangible output of research and development work undertaken by an individual. They also provide a means for a broad international comparison between various nations. USA, countries of European Union, and other developed countries produce the majority of these publications. China is now fast catching up with them with an unprecedented growth rate in the recent years. The encouraging part, however, is that developing countries like India and Brazil have also experienced rapid growth in publication of research articles. This is indicative of positive growth of research and development-related jobs in the country. As indicated by NIRF in their report (NIRF, 2017), it is observed that Indian share of the overall world publications is merely about 3.51% (Table 2). In case of Engineering category, Indian share is around 5.43% which is also not

Table 2 Research publications of NIRF institutes in comparison to total research publications of the World and India (NIRF, 2017)

Discipline/category	No. of research publications		
	World (1)	India (2)	NIRF eligible institutes
Overall (all)	7,491,367	263,125 3.51% of (1)	178,693 67.91% of (2)
Engineering	2,241,598	121,615 5.43% of (1)	82,507 67.84% of (2)
Management	94,113	1704 1.81% of (1)	701 41.14% of (2)
Pharmacy	193,580	8593 4.44% of (1)	3046 35.45% of (2)

something to be excited about. Nearly 68% of the scholarly output from India is represented in the NIRF evaluation. Hence, it can be safely concluded that this would be close to (if not equal to) the total scholarly output from India. When comparison is made between the top 100 ranked (based on research publications) engineering institutions in India, IITs take the lion's share with nearly 35% of the Engineering publications to their credit (Table 3). What may be appreciated, however, is the fact that NITs and some of the State universities and deemed-to-be-universities also make significant contributions to our Engineering publications' share. This clearly augurs well for the research productivity in the engineering domain. Fig. 9 shows that NIRF applicants occupy about 67.84% of the total engineering publications from India. The unrepresented part (32.16%) mainly consist of the publications of Research Labs, belonging to the CSIR System,

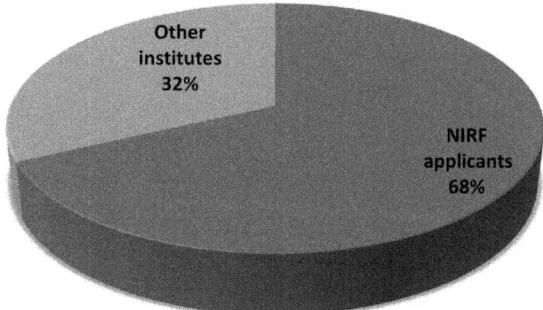

Fig. 9 Percentage share in India's total publications vs publications of NIRF applicants in Engineering (NIRF, 2017).

Table 3 Share of publications from top 100 highly publishing engineering institutes (NIRF, 2017)

Type of institute	No. of publications	Percentage share
IIT'S	23,018	35.4
NIT'S	9854	15.15
State Universities	10,064	15.45
Deemed Universities	7843	12.06
Private Universities	2131	3.28
Colleges	4557	7.01
Other CFTIs (Central-funded technical institutes)	1947	2.99
Others	5613	8.63

the DAE system, DRDO system, ISRO, and Private Research Labs who also publish significantly in Engineering and do not form part of the mainstream academic system. Hence, the research contribution of the third grade institutions (non-applicants) is minuscular.

7.5.1 Availability of Fellowships for Carrying Out Research in Engineering

In order to motivate the candidates with sound technical knowledge and who want to carry out research at Doctoral level, various financial assistances are provided by both the Central Government and the State Governments. In case of Central Government-funded institutes, fellowship is provided by the respective institute to the student selected for Doctoral research.

Both UGC and AICTE offer various fellowships for carrying out research works. National Doctoral Fellowship, Rajiv Gandhi National Fellowship, Maulana Azad National Fellowship, etc. are some of the fellowships offered by them. Apart from these, Council of Scientific and Industrial Research (CSIR) also provide fellowship for carrying out research work in various streams of engineering. Besides, fellowships are also offered from time to time by Department of Science and Technology (DST), viz., DST-INSPIRE, etc. Women-specific scholarships are also given by DST to promote gender equality in case of higher education and research. For majority of the schemes, qualifying candidates are normally eligible to receive a monthly fellowship for a maximum duration of 5 years. The fellowship amount is in general sufficient for supporting the day-to-day expenses of a candidate in any Tier 1 city in India. House rent allowance is also paid over and above the fellowship amount. In addition, a scholar receives an annual contingency grant to meet expenses, viz., buying books, computer peripherals, and attending conferences/workshops. Apart from the regular fellowships, agencies, viz. ISRO, DRDO, etc., also outsource various research projects to reputed institutions in which provision of scholarships for a few candidates is present. A nominal stipend is also provided for PG students (who qualify in GATE examination) in technical branches by Ministry of Human Resources and Development (MHRD), Government of India. Although a variety of fellowships are offered for pursuing doctoral research, their number is still insufficient considering the huge population of India. For example, in a given year, UGC typically opens 8000 fellowship slots for Indian citizens (Reddy et al., 2016). Due to unavailability of scholarships, many interested candidates are forced to leave the course midway, while some continue by spending from their own pocket.

8 RECOMMENDATIONS FOR THE DEVELOPMENT OF MANUFACTURING ENGINEERING EDUCATION IN INDIA

8.1 Larger Involvement of Public Sector in Technical Education

With the advent of liberalization and privatization of the economy, background was set for increased growth of the country's economy. This prompted private initiative to accelerate the pace of technical education in India. As a result, there was a spree of setting up of engineering colleges in the country. However, the motive behind the involvement of the private sector in higher education, particularly in engineering and technical education, has changed from philanthropy to profit with the advent of new liberal regulations and market policies (Choudhury, 2016). This led the technical education sector in India drift toward a commercial business activity. It is unfortunate that even after repeated warnings from the Government to the private sector against commercialization of the educational services provided by them, it is being neglected.

Now, privatization and commercialization of technical education being two sides of the same coin, it is difficult to distinguish between these two either theoretically or empirically (Choudhury, 2016). Both are based on the same principle of making and maximizing profits. Thus, privatization will ultimately lead to commercialization. Therefore, emphasis should be laid on the larger involvement of the public sector in technical education and also in regulating the expansion of the private sector. Colleges identified as fake or with substandard infrastructure should be shut down and their license canceled. Also, regular auditing and monitoring of the activities of the private colleges would enable prevent too much degradation in the quality of education they deliver.

8.2 Introduction of MOOCs for Engineering Education in India

MOOCs is the abbreviated form of Massive Open Online Courses. MOOCs offer online courses in a variety of subjects with the help of internet, which people can join from anywhere in the world. Most of the reputed Universities in the world are joining this noble venture (Joshi, 2015). A majority of these courses are taught by renowned teachers and are offered free of cost. Hence, one can get access to quality education with minimal cost by joining MOOCs.

Universities are motivated to put their courses online by setting up open learning platforms such as edX. Coursera and Udacity platforms are working for commercial purpose with prestigious world Universities. They offer online courses for free or charge nominal fee for certification and credit for the award (Joshi, 2015). Government of India, through MHRD and with the help of the IITs and IISc, has already launched free online course name NPTEL (National Programme on Technology Enhanced Learning) about 15 years back. IIT Bombay and IIT Delhi have also started their online courses.

MOOCs can be an answer to various challenges in engineering education which a country like India is currently facing. Problems like unqualified faculty, outdated curriculum, etc., can be addressed to a large extent by accepting MOOCs which are reliable to a great extent. Students can learn from the world class faculties through internet facility, only which nowadays is easily accessible on smartphones. Even students from rural or poor background can gain a lot of technical knowledge through these courses. These courses can also be used for doubt clearance and supplementing the class teachings. In case of interest in a particular subject which may not be present in the curriculum, a student can refer to MOOCs.

However, MOOCs have their own limitations. MOOCs cannot replace personal teaching experiences, which many times are more effective in communicating certain concepts especially for application-oriented technical courses. Besides, laboratory experiences and field work performed under the observation of faculty cannot be replaced by MOOCs. Hence, there is a recommendation of blended education with MOOCs in technical courses (Joshi, 2015). Blended MOOCs can be more powerful than conventional teaching techniques. According to a proposed model by IIT Bombay, in blended MOOCs, students can study available digital material on their own pace by viewing prerecorded lectures. After that, they attend classroom tutorials and discussion sessions. Finally, students solve problems in groups under the supervision of a faculty.

8.3 Maintaining and Monitoring of Quality of Education in Tier-2 and Tier-3 Colleges

The majority of the engineering colleges in India fall under Tier-2 and Tier-3 categories. Hence, ensuring the quality of education for these two category colleges would lead to the enhancement of the overall engineering education scenario in India.

8.3.1 Implementation of Outcome-Based Education in Engineering Institutes

Outcome-based education (OBE) is student-centered instruction model that focuses on measuring student performance through outcomes.

Outcomes include attributes, viz., knowledge, skills, and attitudes. Its focus remains on evaluation of outcomes of the program by stating the knowledge, skill, and behavior a graduate is expected to attain upon completion of a program and after 4–5 years of graduation. In the OBE model, the required knowledge and skill sets for a particular engineering degree are predetermined and the students are evaluated for all the required parameters (Outcomes) during the course of the program.

In OBE, the focus is on Outcomes which creates a clear expectation of what needs to be accomplished by the end of the course. This enables the students to understand what is expected of them and also the teacher to know what they exactly need to teach during the course. Instructors will be able to structure the lessons according to the student's needs. OBE provides a tangible measure of the teaching-learning process which was previously limited only to the result of the students in final examination. OBE assesses the teaching-learning ecosystem in all the aspects and gives a more practical statistics which can then be compared across different institutions in India.

OBE-based education has been implemented in many of the developed nations of the world, viz., the United States, Australia, Canada, Japan, Korea, the United Kingdom, etc., and the results are quite encouraging. Such a system, if implemented in the engineering institutes in India, would surely maintain a quality technical education across the nation.

8.3.2 Mandatory Accreditation for the Central and State Universities and Private Engineering Colleges

Assessment and accreditation in the higher education through a transparent and informed external review process are an effective means of quality assurance (UGC, 2016). It also provides a common frame of reference for students to obtain credible information on academic quality. This also assists in student migration across institutions, domestic as well as international.

Assessment is undertaken prior to the commencement of academic programs in an Institution, whereas accreditation is undertaken after an institution attains certain years of existence (6 years) or passing out of specified number of batches (two batches), whichever is earlier (UGC, 2016). NAAC (National Assessment and Accreditation Council) and NBA (National Board of Accreditation) are two bodies which currently provide assessment and accreditation of institutes of higher education in India. NAAC accredits general colleges and universities, while NBA accredits technical programs, such as engineering and management programs.

The importance of assessment and accreditation is increasingly understood by the Indian Government and UGC is pressurizing the colleges and universities to get accredited. According to a recent decision made by UGC (UGC, 2016), "No Higher Educational Institution or its Faculties, Schools, Departments, Centers or any other units therein, by whatever name called, shall be eligible for applying or receiving financial assistance from the Commission from 1st April, 2016 onwards, under any of its schemes without having undergone assessment and accreditation on or before 31st December, 2015." UGC has also reported that out of 40 Central Universities, 24 Central Universities have obtained the NAAC accreditation which is an encouraging figure.

The induction of India in the Washington Accord in 2014 with the permanent signatory status of National Board of Accreditation (NBA) is considered a big leap forward for the higher-education, especially technical education system in India. It implies that an engineering graduate from India can be employed in any one of the other countries who have signed the accord. For Indian engineering institutions to get accredited by NBA according to the pacts of the accord, it is compulsory that engineering institutions follow the OBE model. Some of the institutions have already obtained accreditation, while many are in the process of applying. The adoption of OBE at engineering institutions is considered to be a great step forward for higher education in India, but the actual success lies in the effective adoption and stringent accreditation process to ensure that the quality of education is maintained.

8.4 Providing Necessary Funding to the Colleges

Lack of funding is one of the burning issues in the higher education sector in India. Although Central Government-funded engineering institutions enjoy a good amount of funding, the same is not true in majority of the State-funded ones. Lack of maintenance (both buildings and equipment), lack of teaching aids, etc., represent the picture of dwindling fund in these colleges. In many State Universities, research works get hampered due to lack of funds. Manufacturing engineering education requires a lot of heavy machine tools. A majority of those remain in broken condition with no resources to repair them or buy a new one. Advanced CNC-based machine tools also could not be installed in many of the institutions due to lack of funds.

UGC provides financial assistance to Universities and colleges under various schemes. Under one of the schemes of General Development Assistance to Colleges, XII plan (UGC, 2016), financial assistance is provided to the colleges for strengthening basic infrastructure and meeting their basic needs like books and journals (including Book Banks), scientific equipment, campus development, teaching aids which are needed for proper instruction, extension/renovation of existing buildings and construction of new buildings, extension activities, facilities for women, etc. However, considering the number of institutions in India, it is needless to say that the amount of funding received is insufficient in most of the cases.

8.5 Regular Revision of Curriculum With Feedback From Industry

To make its syllabus up-to-date and industry-relevant, Universities should be proactive to implement what is contemporary and in practice. Universities can always do a better job of keeping track of the tools used in industry and apply those tools in the basic learning. For example, cybersecurity is now a core competency in the performance evaluation and all the machines in a shop are connected to the cloud. Hence, the worker must understand the cybersecurity protocols. Manufacturing is one of the most highly hacked industries and that is where lot of intellectual property which is very critical to the national security of a country should be safe and guarded.

For revising syllabus and curriculum, regular interaction between the engineers from the industry and the faculty members may be arranged. A student trained in an updated curriculum is more acquainted with modern techniques which increases his probability of being employed after the completion of the course.

8.6 Training and Development of Teachers

As already pointed out, lack of qualified teachers degrades the quality of engineering education as a whole. Hence, teachers with graduate degrees are to be encouraged to get M.Tech. or M.E. degrees, whereas teachers with PG degrees are to be encouraged to pursue Doctoral degrees. Necessary incentives could be granted for this purpose. Moreover, seasoned teachers are adverse to the idea of change and they continue to embrace outdated teaching methodologies and promote rote learning, which deters conceptual understanding and creative thinking in students. Hence, there is a need to conduct faculty development programs time to time. Faculties should be encouraged to attend training courses which may change the outlook of the way of teaching they are currently involved in and which will finally benefit the students. Pedagogical courses are increasingly becoming popular for training the teachers. The UGC has been making proactive efforts to upgrade the knowledge and skills of faculty members in the institutions of higher education. For the purpose of organizing orientation and refresher courses for in-service faculty members, the UGC has established and funds a network of 66 Academic Staff Colleges across the country.

When manufacturing engineering education is concerned, the industrial experience of the faculty members matters a lot in their teaching methodology. In many of the developed nations in the world, engineering faculty members spend a good amount of time in the industry regularly. This gives them the necessary practical knowhow as well as keeps them updated with the latest technologies adopted worldwide. Finally, when they teach in the colleges, they are able to create the necessary enthusiasm among the students which strengthens this teaching-learning relationship. Such a mechanism of regular industrial exposure to the engineering faculty members in India will definitely enrich the manufacturing engineering education. However, to introduce such a mechanism, strong relations between industry and academia must be developed and regular interactions together with joint project must be taken up.

8.7 Project-Based Learning Linked With Industry

For advancement in the manufacturing engineering education, project-based learning linked with the industry has become a necessity. Students can be given simple real-life industrial problems which they would attempt to solve on their own under the supervision of a faculty. In case of clarifications, students would interact with the industrial engineers who would also guide them toward the solution. In this way, student would acquaint themselves with the industrial culture and have a practical outlook over the engineering problems. Besides, students would also be aware of the industrial practices which ultimately would help them grow as an engineer. For example, mechanical handling in industry, although taught as a subject in some institutes, is actually a thing to experience. How a large and heavy component is moved from one place to another and assembled with another is a practical problem which is impossible to appreciate in the classroom teaching. Handling of real-life problems would also create enthusiasm among students toward positive learning.

9 INITIATIVES FOR THE DEVELOPMENT OF MANUFACTURING ENGINEERING EDUCATION

Ministry of Human Resource Development, Government of India, has been quite proactive with many initiatives to improve the engineering education in the country. Some of them are discussed in the following texts.

9.1 Launch of SWAYAM Platform and Engineering MOOCs in India

Government of India realizes the importance of online learning, and hence, a few MOOCs have been launched in India. SWAYAM is one such notable program initiated by Government of India and designed to achieve the "three cardinal principles of Education Policy viz., access, equity and quality" (MHRD-GoI, 2017). The objective of this effort is to take the best teaching-learning resources to all, including the most disadvantaged. SWAYAM seeks to bridge the digital divide for students who have hitherto remained untouched by the digital revolution and have not been able to join the mainstream of the knowledge economy (MHRD-GoI, 2017).

SWAYAM is hosted on an indigenous IT platform which provides access to courses, taught in classrooms from 9th class till postgraduation, to anyone anywhere at any time. All the courses are interactive, prepared by the best teachers in the country, and are available free of cost to the residents in India. More than 1000 specially chosen faculty and teachers from across the Country have participated in preparing these courses (MHRD-GoI, 2017). The courses hosted on SWAYAM are divided into 4 segments—(1) video lecture, (2) specially prepared reading material that can be downloaded/printed, (3) self-assessment tests through tests and quizzes, and (4) an online discussion forum for clearing the doubts. Steps have been taken to enrich the learning experience by using audio-video and multimedia and state of the art pedagogy/technology.

SWAYAM platform is indigenously developed by MHRD and AICTE with the help of Microsoft and would be ultimately capable of hosting 2000 courses and 80,000 h of learning: covering school, undergraduate, postgraduate, engineering, law, and other professional courses (MHRD-GoI, 2017). In case of engineering, already 236 courses are available among which many courses of manufacturing engineering are also present. Hence, SWAYAM has all the potentials to be responsible for the enhancement of engineering education in country.

India already had an online learning platform even before MOOCs became popular. The open learning content is offered by the IITs and IISc through NPTEL as already mentioned in one of the previous subsections. NPTEL has started online course in computer science, electrical engineering, mechanical engineering, ocean engineering, management, humanities, music, etc. (Joshi, 2015). They also grant certificate for particular course.

9.2 Funding of Engineering Colleges Under the Umbrella of TEQIP

Government of India implemented TEQIP (Technical Education Quality Improvement Programme) project to improve the quality of technical education system in the country (AICTE, 2017b). During the period 1991–2007, NPIU (National Project Implementation Unit) implemented three technical education Projects of Government of India assisted by World Bank, which helped to strengthen and upgrade the technical education System and benefited 552 polytechnics in 27 states (AICTE, 2017b). World Bank has rated these projects highest based on Project Management and Implementation. Driven by the success of these projects, Government of India sought similar financial assistance from World Bank for a systemic transformation of the technical education system as a whole with special focus on overall quality improvement in engineering education. World Bank reciprocated positively by showing keen interest to make India's technical education system globally competitive and showed its willingness to assist the Government of India to launch a Technical Education Quality Improvement Programme (TEQIP) as a long-term programme of 10–12 years and in 2 or 3 phases. TEQIP has successfully completed the second phase with the third phase on the verge of implementation. It is relevant to put up here the objectives of the various TEQIP phases so that its implication on the overall technical education can be realized and appreciated.

TEQIP Phase-I

Implemented in 13 States and covered 127 Institutions including 18 Central Government-funded Institutions.

Objectives of TEQIP Phase —I (AICTE, 2017b)

- Promotion of Academic Excellence
- System Management Capacity Improvement
- Networking of Institutions for quality enhancement and resource sharing
- Enhancing quality and reach of services to Community and Economy

TEQIP Phase-II

Implemented in around 200 engineering institutions.

Objectives of TEQIP Phase–II (NPIU, 2017a)
- Strengthening institutions to improve learning outcomes and employability of graduates
- Scaling-up Postgraduate Education and demand-driven Research & Development and Innovation
- Establishing Centers of Excellence for focused applicable research
- Training of faculty for effective teaching, and
- Enhancing Institutional and System Management effectiveness

TEQIP Phase-III
Objectives of TEQIP Phase–III (NPIU, 2017b)
- Improving quality and equity in engineering institutions in focus states, viz., seven Low Income States (LIS), eight states in the North-East of India, three Hill states, viz., Himachal Pradesh, Jammu & Kashmir, and Uttarakhand, and Andaman and Nicobar Islands (a union territory (UT))
- System-level initiatives to strengthen sector governance and performance which include widening the scope of Affiliating Technical Universities (ATUs) to improve their policy and academic and management practices towards affiliated institutions
- Twinning Arrangements to Build Capacity and Improve Performance of institutions and ATUs participating in focus states

9.3 Implementation of GIAN Courses

GIAN or Global Initiative of Academic Networks is Government of India-approved program which enabled foreign academicians to come and teach in India at government institutions. However, any student or faculty in India can register for his/her choice of course and attend the same. The aim of this scheme is to tap the talented pool of scientists and entrepreneurs, internationally, to encourage their engagement with the institutes of Higher Education in India so as to augment the country's existing academic resources, accelerate the pace of quality reform, and elevate India's scientific and technological capacity to global excellence. The experts would share their experiences and expertise to motivate people to work on Indian problems.

From manufacturing engineering point of view, courses, viz., Smart Manufacturing, Advanced Process Planning, etc., are listed under GIAN-approved courses (GIAN, 2015). Apart from these, many advanced and interdisciplinary topics have also been or are going to be covered. Hence, this type of initiative would surely have a positive impact on the manufacturing engineering education of the nation.

10 PROMOTION OF INDIAN MANUFACTURING SECTOR

In the last 5 years, Government of India has taken some steps to boost the manufacturing sector in India. Programs, viz., Make in India and Skill India, have been launched in a wide scale. This has created a positive stir in the manufacturing community in India towards nation building through industrialization. Overall, there is also a positive outlook towards the prospects of manufacturing engineering education for fetching good jobs.

10.1 Make in India Initiative: Emphasis on Indigenous Manufacturing

Make in India is one of the flagship campaigns of Indian Government intended to boost the domestic manufacturing industry and attract foreign investors to invest into the Indian economy. This is part of wider set of nation–building initiatives. Launched in 2014, Make in India is devised to transform India into a global design and manufacturing hub (GoI, 2014). It is launched with the intention of reviving manufacturing businesses and emphasizing key sectors in India amidst growing concerns that most entrepreneurs are moving out of the country due to its low ratings in ease of doing business.

Manufacturing currently contributes just over 15% to the Indian GDP. The aim of this campaign is to grow this to a 25% contribution as seen with other developing nations of Asia. In the process, the government expects to generate jobs, attract much foreign direct investment, and transform India into a manufacturing hub preferred around the globe.

The Make in India initiative is based on four pillars, which have been identified to give boost to entrepreneurship in India, not only in manufacturing but also other sectors (PMO, 2017).

- New Processes

    Make in India recognizes "ease of doing business" as the single most important factor to promote entrepreneurship. A number of initiatives have already been undertaken to ease business environment. The aim is to de-license and deregulate the industry during the entire life cycle of a business.

- New Infrastructure

    Availability of modern and facilitating infrastructure is a very important requirement for the growth of industry. Government intends to develop industrial corridors and smart cities to provide infrastructure based on state-of-the-art technology with modern high-speed communication and integrated logistic arrangements. Existing infrastructure is to be strengthened through upgradation of infrastructure in industrial clusters. Innovation and research activities are supported through fast paced registration system and, accordingly, infrastructure of Intellectual Property Rights registration setup has been upgràded. The requirement of skills for industry are to be identified and accordingly development of workforce to be taken up.

• New Sectors:

Make in India has identified 25 sectors in manufacturing, infrastructure, and service activities and detailed information is being shared through interactive web-portal and professionally developed brochures. FDI has been opened up in Defense Production, Construction, and Railway infrastructure in a big way.

• New Mindset:

Industry is accustomed to see Government as a regulator. Make in India intends to change this by bringing a paradigm shift in how Government interacts with industry. The Government will partner industry in economic development of the country. The approach will be that of a facilitator and not regulator.

Based on Make in India wave, six industrial corridors are being developed across various regions of the country. Industrial Cities will also come up along these corridors. India received Rs. 16.40 lakh crore (US$260 billion) worth of investment commitments and investment inquiries worth Rs. 1.50 lakh crore (US$23 billion) between September 2014 and February 2016 have been made (PTI, 2016). Due to this initiative, there is visible momentum, energy, and optimism in the education sector as well. There is an effort to create a holistic strategy for education and workforce development in manufacturing.

10.2 Skill India Mission

Skill India is another initiative of Government of India whose motive is to empower the youth of the country with skill sets which make them more employable and more productive in their work environment (GoI, 2015).

India has one of the largest youth populations in the world with 65% of its youth in the working age group (GoI, 2015). In order to tap this demographic advantage, skill development of the youth is necessary so that they add not only to their personal growth, but to the country's economic growth as well. Skill India offers training courses covering 40 sectors in the country which are aligned to the standards recognized by both, the industry and the government under the National Skill Qualification Framework (GoI, 2015). The courses enable a person focus on practical delivery of work and help him/her enhance his technical expertise so that he/she is ready from day one of his/her job and companies do not have to invest into training him/her according to his/her job responsibilities.

This mission was launched in 2015 and targets to train more than a crore fresh entrants into the Indian workforce have been substantially achieved for the first time. 1.04 crore Indians were trained through Central Government Programs and NSDC (National Skill Development Corporation) associated training partners in the private sector (GoI, 2015).

The success of a nation always depends on the success of its youth and Skill India is certain to bring a lot of advantage and opportunities for these young Indians. The time is not far when India will evolve into a skilled society where there is prosperity and dignity for all (GoI, 2015).

11 ECOSUSTAINABLE AWARENESS IN MANUFACTURING ENGINEERING EDUCATION

Higher contribution from the manufacturing sector is required for the development of any country. However, enhancement in the manufacturing sector has been accompanied by growing global environmental concerns, such as climate change, energy security, and increasing scarcity of resources (OECD, 2009). Over the last few decades, the environmental burden linked to industrial activities has become an increasingly important global issue and a great challenge for the society (Pathak et al., 2017). In response, manufacturing sector has recently shown more interest in sustainable manufacturing and has adopted certain corporate social responsibility initiatives (OECD, 2009). However, such efforts fall far short of meeting these pressing challenges. Moreover, improved efficiency in some regions has been offset by increases in consumption and growth in others.

Due to its widespread implications, environmental issues have already cropped up in the Indian secondary educations curriculum in the form of Environmental Science-related subjects. However, in case of engineering, education is mostly based on teaching of techniques to achieve a particular target or provide solution to a particular application. The environmental concern/issue related to a particular engineering application is sidelined. This is quite a serious concern as the students especially belonging to disciplines, viz., Mechanical Engineering, Manufacturing Engineering, etc., who would be in future joining the industrial workforce, are not trained about ecosustainability at the very grass root level. Students are not proficient in the areas, viz., resource availability and economics, reuse and recycling options, life-cycle assessment, ecoindustrial development, etc. Hence, the current scenario in which the nation is trying to make its mark in the global manufacturing sector, the negligence of ecosustainable topics in the engineering curriculum is of a great concern. It is high time that the regulatory bodies of higher education system in India take necessary action about this.

12 BRIEF COMPARISON OF MANUFACTURING EDUCATION WITH DEVELOPED COUNTRIES

It is thought to be quite relevant to bring in a comparison of Indian manufacturing engineering education with that of other developed nations of the world. This way, one can get an idea about the relative position of Indian system globally and what steps can be taken to improve the same. Initially, the scope and availability of such course in various nations as discussed by Khare et al. (2016) is presented. Khare et al. (2016) considered QS university ratings and the BRICS countries for this comparison. They reported that Tsinghua University, China, has a separate Institute for Manufacturing Engineering. It was established in 1996 and has much application-based research centers focusing on computer-integrated manufacturing, robotics, graphics, and CAD. Shanghai Jiao Tong University has a department for Industrial Engineering. Other top universities of China do not have separate departments for manufacturing related courses. In case of Russia, Bauman Moscow State Technical University of Russia has a program in Industrial engineering. Similar to China, no manufacturing-oriented courses are present in the other top institutions in Russia. Khare et al. (2016) reported that University of São Paulo, Brazil, offers many graduate programs including manufacturing engineering. This is due to the Brazilian government which has provided a vision for education prioritizing on science and technology and which also reflects in their engineering curriculum. They further reported that Brazil and South Africa have a better laid out plan for expansion of manufacturing engineering courses. Surprisingly, Russia has a few colleges focusing on industrial or manufacturing engineering. According to the observation of Khare et al. (2016), India stands at par with China with respect to manufacturing engineering education curriculum.

Although from the above discussion it is found that India maintains a decent position in terms of manufacturing education among the BRICS countries, India is beyond doubt severely lagging behind in comparison to the top developed nations of the world, viz., United States, Germany, Japan, etc. Interestingly, United States Department of Defense (DoD) recently (in 2017) launched the "Manufacturing Engineering Education Grant Program" which aims to help strengthen the U.S. economy and national security, while safeguarding the competitiveness of the U.S. manufacturing sector. According to ASME official webpage (ASME, 2017), "The new program has great potential to strengthen national security and increase economic competitiveness by improving and modernizing the U.S. industrial base. Through this program, students, technologists, and manufactures will be better equipped to manufacture U.S. military equipment and technology domestically, protecting and securing the future of the American Warfighter. The Manufacturing Engineering Education Grant Program is intended to not only strengthen the U.S. military's capabilities, but also allow the United States to compete against other nations economically." Some of the objectives that the institutions need to fulfill through this program are:

- Curriculum development
- Fostering interactions between student and industry
- Job placement activities
- Establishing joint manufacturing programs with space and defense laboratories

Such program has great potential to strengthen the national economic security of any country by working to bridge the skills gap that we hear about so often in our society as we continue to advance and innovate into the 21st century.

13 CLOSURE

An improved manufacturing engineering education gives rise to quality manufacturing workforce who can participate in nation building by strengthening the manufacturing sector of the country. As GDP of a country is directly related to its expanded manufacturing capability, a strong manufacturing engineering education is a great need of the hour especially for a developing country like India. Through proper nurture and care of the future manufacturing workforce only, India can dream to be at par with the developed nations of world in the near future.

REFERENCES

AICTE, 2017a. https://www.aicte-india.org/about-us/overview. (Accessed 5 December 2017).

AICTE, 2017b. https://www.aicte-india.org/bureaus/rifd/teqip. (Accessed 6 December 2017).

ASME, 2017. https://www.asme.org/about-asme/news/asme-news/congressional-briefing-highlights-dods. (Accessed 8 December 2017).

Choudhury, P.K., 2016. Growth of engineering education in India: status, issues and challenges. Higher Educ. Future 3 (1), 93–107.

GIAN, 2015. http://www.gian.iitkgp.ac.in/ccourses/approvecourses3. (Accessed 15 February 2017).

GoI, 2014. http://www.makeinindia.com/about. (Accessed 18 December 2017).

GoI, 2015. http://www.skilldevelopment.gov.in/background.html. (Accessed 11 December 2017).

GoI, 2016. All India survey on higher education (2015–16). http://mhrd.gov.in/sites/upload_files/mhrd/files/statistics/AISHE2015-16.pdf. (Accessed 10 December 2017).

Indian Education, 2017. http://www.indiaeducation.net/careercenter/engineering/manufacturingengineering/manufacturing-engineering.aspx. (Accessed 14 December 2017).

Jain, D., Nair, K.S., Jain, V., 2015. Factors affecting GDP (manufacturing, services, industry): an Indian perspective. Ann. Res. J. SCMS Pune 3, 19.

Joshi, N.M., 2015. Acceptance of MOOCs for engineering education in India. 10th International CALIBER-2015. HP University and IIAS, Shimla, Himachal Pradesh, India, INFLIBNET Centre, Gandhinagar, Gujarat, India, paper ID 98.

Khare, S., Chatterjee, A., Bajpai, S., Bharati, P.K., 2016. Manufacturing engineering education in India. Manage. Prod. Eng. Rev. 7, 40.

MHRD-GoI, 2017. https://swayam.gov.in/about. (Accessed 15 December 2017).

NASSCOM-McKinsey, 2011. NASSCOM perspective 2020: transform business, transform India. http://mhrd.gov.in/sites/upload_files/mhrd/files/statistics/AISHE2015-16.pdf. (Accessed 10 December 2017).

National Science Board, 2016. Science and engineering indicators 2016. Arlington, VA, National Science Foundation. NSB-2016-1.

NIRF India Rankings 2017, 2017. https://nirfcdn.azureedge.net/rankingpdf2017/IR2017_Report.pdf. (Accessed 11 December 2017).

NPIU, 2017a. TEQIP II. http://www.npiu.nic.in/teqip2introduction.html. (Accessed 8 December 2017).

NPIU, 2017b. TEQIP III: project objectives. http://www.npiu.nic.in/teqip3projectobjective.html. (Accessed 8 December 2017).

OECD, 2009. Sustainable manufacturing and eco-innovation: framework. Practices and Measurement-Synthesis Report. Parishttps://www.oecd.org/innovation/inno/43423689.pdf. (Accessed 17 December 2017).

Pathak, P., Singh, M.P., Sharma, P., 2017. Sustainable manufacturing: an innovation and need for future. International Conference on Recent Innovations in Engineering and Technology, Jaipur, India, Maharishi Arvind College of Engineering & Research Centre. ISBN 978-93-86291-63-9.

Peters, J., 1989. Manufacturing in mechanical engineering education in developing countries. Eur. J. Eng. Educ. 14 (2), 135–139.

PMO, 2017. PM INDIA-major initiatives. http://www.pmindia.gov.in/en/major_initiatives/make-in-india/. (Accessed 20 December 2017).

PTI, 2016. Make in India Week gets Rs 15.2 lakh crore investment commitments. https://economictimes.indiatimes.com/news/economy/policy/make-in-india-week-gets-rs-15-2-lakh-crore-investment-commitments/articleshow/51040369.cms. (Accessed 17 December 2017).

Reddy, K.S., Xie, E., Tang, Q., 2016. Higher education, high-impact research, and world university rankings: a case of India and comparison with China. Pacific Sci. Rev. B Human. Social Sci. 2 (1), 1–21.

Revathy, L.N., 2017. Manufacturing Sector Should Grow, Contribute 25% to GDP: Expert. The Hindu Business Line, Coimbatore, The Hindu Group. http://www.thehindubusinessline.com/economy/manufacturing-sector-should-grow-contribute-25-to-gdp-expert/article9455444.ece. (Accessed 5 December 2017).

UGC, 2016. Annual report 2015–16. (New Delhi). https://www.ugc.ac.in/pdfnews/3710331_Annual-Report-2015-16.pdf. (Accessed 14 December 2017).

UGC, 2017. https://www.ugc.ac.in/page/Genesis.aspx#. (Accessed 14 December 2017).

Yashpal Committee Report, 2008. Report of the committee to advice on renovation and rejuvenation of higher education. http://www.aicte-india.org/downloads/Yashpal-committee-report.pdf#toolbar=0. (Accessed 15 December 2017).

CHAPTER 3

Learning Enhancement of Project-Based Unit in Mechanical Engineering Undergraduate Course

A. Pramanik*, M.N. Islam*, A.K. Basak[†], Y. Dong*
*School of Civil and Mechanical Engineering, Curtin University, Bentley, WA, Australia
[†]Adelaide Microscopy, The University of Adelaide, Adelaide, SA, Australia

1 INTRODUCTION

Manufacturing, a complex system, is compose of production processes and systems that involve manpower and equipment (Todd et al., 2001). To keep pace with increasing demand of modern global economy, the size and overall scope of manufacturing process became more and more complex. In recent time, manufacturing industries around the globe have undergone dramatic changes as a result of the effects of industrial globalization. One obvious impact has been the outsourcing of manufacturing to keep products competitive in the global market. These trends, currently redefining the manufacturing enterprises, now serve as motivators for major changes in academia (Ziemian and Sharma, 2008). To keep the manufacturing process economically viable and efficient, successful manufacturers constantly analyze the market and take necessary steps toward redesign and upgradation of technologies, infrastructure, business processes, supply chains, marketing strategies, and overall customer relations (Moon, 2004). With rapid development in new manufacturing processes, such as 3D printing, rapid prototyping, direct digital printing, additive manufacturing, etc., manufacturing industries as well as its supportive learning curricula stand in a turning theme of manufacturing history. Development of such advanced and low-cost techniques are so rapid that, in the last 10–15 years, manufacturing industries have to develop a number of new vocabulary to address the new process and differentiate them from old ones (Flynn, 2012). To take the benefits of such new manufacturing processes on board, the constituents of a manufacturing arrangement (e.g., design, administration, procuring, and fabrication) are

Manufacturing Engineering Education
https://doi.org/10.1016/B978-0-08-101247-5.00003-4

73

combined to cope recurrent variations in products and methods. To became successful in this highly competitive market and dynamic atmosphere, a fresh engineering university graduate essentially cultivates a complete vision of the entire business procedure from drawing board to distribution (Sackett and McCluney, 1992).

In academia, the study of manufacturing is vital toward preparing fresh graduates in this field and, toward that, related course materials will include the aspects of above-mentioned new manufacturing process. To maintain proper economic growth, significant scientific developments and success in fabrication are the driving force behind that. A fruitful communication among diverse sectors of the society is also foreseen to transform the ideas into meaningful reality and implies that today's worldwide economy is not limited to the traditional workshop floor. These industrial tendencies in sequence necessitate the modern learning organization to be adequately extensive and diverse so that it teaches students to be creative in extremely technical sphere (Bengu and Swart, 1996). In other words, it means that modern manufacturing education needs to be interdisciplinary in nature, compiling the essence of core subjects (Bengu, 1995). However, the implementation of the cooperation of engineering disciplines in manufacturing is absent in the syllabus toward preparing students with a thorough understanding of industrial procedures and systems (Bengu and Swart, 1996). Though it has been proved that hands-on education approaches are idyllic to demonstrate the intricacy of contemporary factories (Dessouky et al., 2001), the exposure of students to real procedures is restricted to instructors' demos or videotapes. There is also the possibility of utilizing simulations, but these do not offer sufficient realism (Council, N. R, 1995). Another method to incredulous these restrictions is to organize cooperative involvements that permit students to work in a industrial facility as a portion of their degree programme (Denton, 1998; Ram et al., 1999), as commonly known as internship. This necessitates additional obligation from pupils as it typically involves a minimum one-semester leave from school. In addition to that, number of spots in the industry to enable such internship is limited and lagging behind with respect to the increasing number of students each year. Part of the problem is also to blame highly automation of modern manufacturing process.

The increasing curiosity and usage of dynamic education module are one of the utmost substantial current advances to address the synergy of multidisciplinary teaching approach in manufacturing courses. Though the conception is not new, expansion and recognition of these tools appear to be at all-time high and growth. Idyllically, dynamic education will adequately involve the students and keep them engaged as a team member toward solving a real problem. A combined three-step instructional conception (Moskowitz and Ward, 1998) was developed to involve dynamic learning, application and, those within and across team, education via documents. This conception of "Do, Apply, and Document" has been used fruitfully in degree and administrative curriculums in manufacturing, along with consulting arrangements with industrialized organizations. In broad, the aim of the approach is to incorporate industrial management and engineering tasks and to set up a uninterrupted multidisciplinary education culture (Moskowitz and Ward, 1998).

Having said these, there are scopes to improve manufacturing education in engineering schools by adding laboratory activities in the curriculum to obtain hands-on experiences. In this study, the students were assigned to manufacture a hammer in the laboratory activities, while completing a unit of manufacturing processes in mechanical engineering graduate program. Thus, it is essential to change the educational paradigm in manufacturing to sharpen the skills of the students as well as innovation process to meet the future challenge of supply and demand of manufacturing (Chryssolouris et al., 2013).

2 THE UNIT "MANUFACTURING PROCESSES (MCEN 2004)"

Manufacturing Processes (MCEN 2004) is a basic unit in undergraduate mechanical engineering course at Curtin University to be educated in second year. Pupils need to acquire and apply analytical ability as well as memorize enormous info associated to manufacturing processes in this subject. The subject involves overview of manufacturing processes and safety, specifications and standards, machining processes and traditional shaping, computer numerical control (CNC) machining, production planning, primary and secondary forming processes, and joining processes. The whole course is divided into 10 classroom lectures and five laboratory activities. The lectures covered the curriculum under various titles, for example, (i) Measurement, inspection, and testing; (ii) Machining processes; (iii) Introduction to CNC machining; (iv) Production, planning, and interchangeable manufacture; (v) Casting processes; (vi) Joining processes; (vii) Forming processes; (viii) Bulk deformation processes 1; (ix) Bulk deformation processes 2; and (x) Sheet metal working. The laboratory activities were on (a) Measurement, (b) CNC programming, (c) Casting, (d) Welding, and (e) Machining project. The course actively makes use of multimedia technology, for example, text, slides, audio–video, and communicating software to make course materials available in diverse formats to avoid monotony. The primary courseware material was arranged by faculty, with sections referred to as topics. An "electronic blackboard" works as a primary interface between course materials and students. The main enthusiasm and learning challenge in planning multimedia courseware is to enhance traditional education techniques and embed vast array of high-technological and instructional materials in it. These include abundant software programs, manuals, and machines. Furthermore, the demonstration needs to be tailor-made in such a way that entry-level pupils are not entirely overawed with enormous info.

The laboratory resources consist of the machines, equipment, and software at the university laboratories to introduce associated manufacturing process principles and outline the details of laboratory activities. Faculties and sufficient numbers of technicians ensure safe working atmosphere for hands-on practices on intricate machinery and simultaneously provide a brief outlook of relevant machineries. It should be noted that such a collaborative environment keeps the students engaged in effective learning process. During this hands-on learning process, students can individually use a computer system, for example, computer-aided manufacturing (CAM), computer numerical control (CNC), computer-aided design (CAD), and independently at their own pace. The video resources were complementary to course materials and proactive practice and the students can access those in their own space either in university or at home. The educational videos associated to the course topics were obtained through educational associations ASME (American society of mechanical engineers) to supplement classroom and laboratory practices.

3 LEARNING OUTCOMES

The course was designed by keeping three learning outcomes in focus. On successful completion, students will be able to (a) Clarify and relate manufacturing processes of engineering materials and modules; (b) Relate the doctrines of primary, secondary, and specialized manufacturing procedures; and (c) Analyze and choose suitable fabrication approaches for particular modules/applications. These learning aftermaths address four graduate attributes: (a) Application of discipline information, (b) Communication expertise, (c) Apply logical skills toward complex problem solving, and (d) self-reliance to explore innovative concepts. The first learning outcome

addresses the first and second graduate element, the second learning outcome addresses the first graduate attribute, and the last learning outcome
addresses the third and fourth graduate element, respectively.

4 THE PROJECT

The students were supplied with raw materials, engineering drawing of a
hammer (Fig. 1), cutting tools, machine tools, machine instructions, safety
instructions, and technician support. The hammer has two parts: hammer
handle and hammer head. Those were manufactured separately and assembled together. The hammer was built step by step throughout the semester.
During manufacturing process, each student has to construct an operation
planning sheet to list all the manufacturing with respect to their sequence
to complete the hammer.

5 THE FEEDBACKS OF STUDENTS ON COMFORT OF LEARNING

Students provided individual commentaries on the module/structure–based
education methodology during the survey. The comments were very positive, some of which are given below:

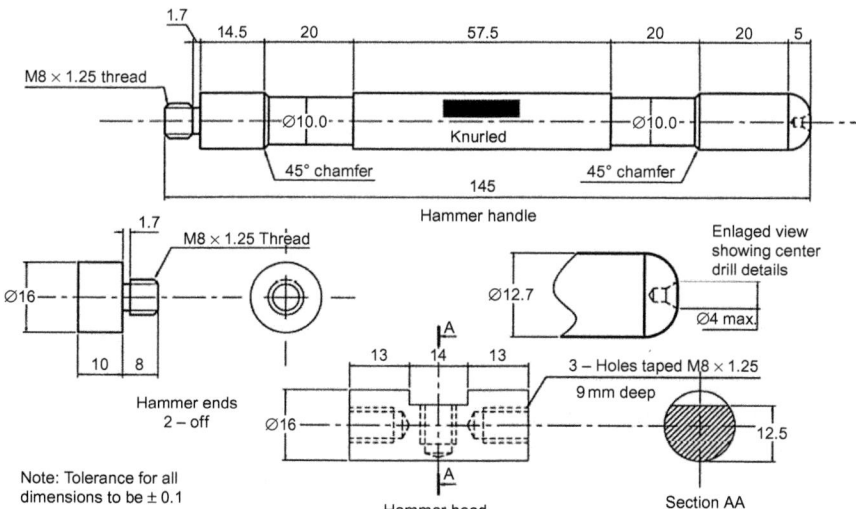

Fig. 1 Engineering drawing of the hammer.

"Thoroughly enjoyed this unit as it is very practical, the lab techs were extremely helpful and friendly and the lectures were very concise and informative. Possibly the best unit i have done so far in the past two years as we were able to take something physical away from what we learnt. Not only did this unit provide vital information for my future career, it also allowed me to develop useful skills which could be helpful when finding part-time or full-time employment".

"The lab part will be the good part to learn manufacturing process. I think conducting the lab for me is the only way to enhance my knowledge, this is because you will have the basics and the general concept, compare to lecture, more virtual".

"Hands on experience rather than a purely death by power point approach was really helpful and enjoyable. The lectures linked to the Workshops well. Exams were well formatted and asked engaging meaningful questions that tested subject knowledge rather than the ability to cram that other information-based (compared to number-based) exams often do. The videos shown in the lectures were a great insight into how what we were learning is applied in the real world".

"The lab component of this unit is outstanding, it engages students and provides them with hands on experience to the concepts they are learning. The laboratory staff are very engaging and helpful".

The above comments clearly show the impact of lab project on students' learning is very encouraging.

6 DISCUSSION

The project provides the students a physical access to get hands-on experience on different machine tools, cutting tools, and measuring devices. The project works keep progressing throughout the semester in addition to other laboratory activities. This gave students enough time to sense industry environment, gain better understanding on the project, and absorb information from class notes to real life problem-solving experience. By using different machine tools, students became familiar with the machines and develop excellent understanding of capabilities and limitations of such machines. The students have to go through different machining operations such as facing, cut-off, turning, chamfering, slotting, milling, tapping, threading, polishing, knurling, and assembling. In addition, they had to deal with given tolerance of final component and holding devices of workpieces and cutting tools. All these resemble to the sense of real manufacturing industry. These were very attractive to the students to keep them motivated and engaged in the project and enable them to finish the project individually with self-confidence and job satisfaction. In addition, students kept the finished product, which also gave them the satisfaction of their achievements.

For the efficient machining, proper sequencing of different machining operations is very important to save time as well as to reduce cost. Faulty products can be fabricated if proper operation sequence is not maintained. It requires logical thinking and overall understanding of all the operations involved and their effects on workpiece geometry. As a result, the students not only get familiar with the machines and machining operations, but also get the opportunity to explore different options and test them based on their creativity. Finally, an operation planning sheet was expected to develop as shown in Table 1.

It is promising to enrich traditional teaching and learning ways by includ- ing diverse technological tools (Pramanik and Islam, 2014a). Technological tools, for example, info and communication tools, virtual games, virtual learning, moveable devices (smart phones and tablets), and others influence teaching and learning methods through controlling learning systems, indi- vidual rejoinder systems, conversation boards, blogs, wikis, societal net- working, podcasts, and a range of web-based applications. In contrast, diverse methodologies in teaching, for instance, critical case investigation, recognizing features of workplace difficulties, etc. necessitate module-based schoolroom teaching. All of these learning tools have been verified for diverse teaching elements in various extents and have enhanced learning capabilities of students (Pramanik, 2013). With technological advancement, the rise of education materials, and interdisciplinary environment of courses, it is the stage to reconsider the approach that engineering themes should be educated to make it real-life-orientated. A number of engineering courses became obsolete with time and loss the appeals among the students and being termed as "problematic and boring" owing to deficiency of appropri- ate upgradation with time and reflect the need of necessity. As a result, pupils decide to give up these courses or become unsuccessful in exams. It is assumed that the dropout rates are not only due to inadequate merit and devotion of pupils as the meritorious students often dropout the course due to lack of interest (Carter and Brickhouse, 1989; Tobias, 1992; Seymour, 1995). This might be reduced by adopting a supportive education methodology, which cares about pupils in terms of inherent enthusiasm, higher-level perceptive, academic and societal support, societal growth, self-confidence, etc. The greatest educational outcome is accomplished once (i) pupils construct on and relate to preceding understandings, (ii) the content is related to them, (iii) there is an opportunity for a

Table 1 Operation planning sheet of a hammer manufacturing process

Op. no:	Description	Machine	Work holding method	Tools used
1	Face one end of hammer head; blank clean	Lathe	3 Jaw chuck	HSS facing tool
2	Face second end of hammer head; blank clean	Lathe	3 Jaw chuck	HSS facing tool
3	Measure blank length	–	–	Vernier calipers
4	Machine blank length to 40 mm ± 0.1 mm	Lathe	3 Jaw Chuck	HSS facing tool/vernier calipers
5	Center drill both ends	Lathe	3 Jaw chuck	HSS center drill
6	Drill Ø 6.8 mm × 10 mm deep; both ends	Lathe	3 Jaw chuck	HSS twist drill 6.8 mm
7	Mill 14 mm × 3.5 mm deep slot	Milling machine	Machine vice & vee block	HSS milling cutter 14 mm
8	Set up in a vee block and mark out center hole	Marking out table	Vee block	Vernier height gauge
9	Center Punch marking out	Vice	Bench Vice	Hammer & center punch
10	Drill Dia 6.8 mm × 10 mm deep	Drill press	Machine Vice & Vee Block	HSS twist drill 6.8 mm
11	Start tap M8 both ends of blank in the lathe	Lathe	3 Jaw chuck	HSS M8 tap
12	Tap middle hole & finish tapping end holes	Bench vice	Bench vice	HSS M8 tap
13	Face one end of hammer handle blank clean	Lathe	3 Jaw chuck	HSS facing tool
14	Face second end of hammer handle blank clean	Lathe	3 Jaw chuck	HSS facing tool

15	Measure blank length	—		Vernier calipers
16	Machine blank length to 145 mm ±0.1 mm	Lathe	3 Jaw chuck	HSS facing tool/vernier calipers
17	Machine step for M8 thread	Lathe	3 Jaw chuck	HSS turning tool
18	Machine radius on the other end	Lathe	3 Jaw chuck	HSS radius form tool
19	Center drill radius end	Lathe	3 Jaw chuck	Center drill
20	Mark out recess area	Marking out table	Vee block	Vernier height gauge
21	Machine recess area	Lathe	3 Jaw chuck & center	HSS chamfering tool
22	Machine thread undercut	Lathe	3 Jaw chuck	HSS undercut tool
23	Machine chamfer on spigot end	Lathe	3 Jaw chuck	HSS chamfering tool
24	Produce M8 thread on spigot end	Lathe	3 Jaw chuck	HSS M8 button die
25	Machine hammer ends to clean	Lathe	3 Jaw chuck	HSS facing tool
26	Knurling hammer handle	Lathe	3 Jaw chuck & center	Knurling tool
27	Polish components	Lathe	3 Jaw chuck	Emery paper
28	Assemble Hammer & fine-tune components			

hands-on involvement, and (iv) pupils can use their individual understanding in cooperation with other pupils and faculty through effective communication (Anthony et al., 1998; Pramanik and Islam, 2014b). The modern engineering profession faces constant challenges like degree of uncertainty, incomplete input data, and challenging (often contradictory) demands from clienteles, managements, ecological groups, and common people. Thus, this necessitates expertise in social relationships along with technological capability to overcome such challenges. Besides incorporation of extra "human" abilities into the knowledge base and professional exercise, the current engineers must also survive with recurrent technical and organizational changes in their workplaces. Additionally, engineers require to get by commercial actualities of industrial practices in the contemporary sphere, along with the legal concerns of each professional choice (Mills and Treagust, 2003). Through the course works and day-to-day activities, students reinforce the above "human" skills in their personality. To complete the training, hands-on experience in any manufacturing unit only can be achieved by arranging extensive laboratory activities where students acquire the perception of complete accountability, creativeness, and enthusiasm of successful accomplishment of a simple manufacturing venture. Consequently, a project-oriented laboratory activity has been considered in this investigation where every student can be assigned to manufacture a metal hammer to address above-mentioned issues. This methodology permits the pupils to gain work-life exposure reasonably independently over prolonged periods and end up with genuine produces or demonstrations (Jones et al., 1997; Thomas, 1999). Real exploration in "design thinking" comprises (i) "convergent component," that produces profound cognitive queries by systematically asking lower-level, (ii) "convergent queries," and (iii) "opposing component" to create the conceptions upon which the "convergent component" may build up (Dym et al., 2005). The education of higher-level cognitive is related to augmented competence of pupils to use those knowledge in problem-solving environments (Thomas, 2000).

7 CONCLUSIONS

The hands-on experience enables the students to apply their engineering knowledge in practical field, taste of practical manufacturing environment as well as gaining research experiences in a specific field. In addition, students feel confident, appreciate creation, are capable of effective thinking toward further improvement of the creation, and apply a wide range of knowledge.

Table 2 Student satisfaction rates due to the introduction of project-based learning

Year	2011	2012	2013	2014
Student satisfaction (%)	81 (283)	90 (208)	93 (246)	97 (242)

The numbers in parentheses are student numbers enrolled in this unit in a particular year.

This also creates a long lasting knowledge in this field in memory. Therefore, in engineering subjects where undergraduate knowledge is applied in day-to-day practical activities, project-based learning units will benefit students as well as industries.

In this study, the student manufactured a hammer by selecting manufacturing processes, developing the sequence of manufacturing processes, and preparing an operation planning sheet. During the process, the students not only get familiar with machines and machining operations, but also get the opportunity to explore different options and explore the options based on their creativity. The students showed a great enthusiasm, creative attitude, and sharp attention, while maintaining a relaxed mode most of the time. The project was introduced in 2012 and improvement of student satisfaction was significant as shown in Table 2, indicating an increase of student satisfaction by 16% in 2014 as opposed to the rate in 2011. This investigation and the above-mentioned key points strongly support project-based laboratory activities with the enhancement of the quality, quantity, and speed of learning.

REFERENCES

Anthony, S.M.H., Spencer, B., Gutwill, J., Kegley, S., Molinaro, M., 1998. The Chem-Links and ModularCHEM consortia: using active and context-based learning to teach students how chemistry is actually done. J. Chem. Educ. 75 (3), 322.

Bengu, G., 1995. Interactive multimedia courseware on manufacturing processes and systems. Int. J. Eng. Educ. 11, 46.

Bengu, G., Swart, W., 1996. A computer-aided, total quality approach to manufacturing education in engineering. IEEE Trans. Educ. 39 (3), 415–422.

Carter, C.S., Brickhouse, N.W., 1989. What makes chemistry difficult? Alternate perceptions. J. Chem. Educ. 66 (3), 223.

Chryssolouris, G., Mavrikios, D., Mourtzis, D., 2013. Manufacturing systems: skills & competencies for the future. Proc. CIRp 7, 17–24.

Council, N. R, 1995. Information Technology for Manufacturing: A Research Agenda. National Academies Press.

Denton, D.D., 1998. Engineering education for the 21st century: challenges and opportunities. J. Eng. Educ. 87 (1), 19–22.

Dessouky, M.M., Verma, S., Bailey, D.E., Rickel, J., 2001. A methodology for developing a web-based factory simulator for manufacturing education. IIE Trans. 33 (3), 167–180.

Dym, C.L., Agogino, A.M., Eris, O., Frey, D.D., Leifer, L.J., 2005. Engineering design thinking, teaching, and learning. J. Eng. Educ. 94 (1), 103–120.

Flynn, E.P., 2012. Design to manufacture—integrating STEM principles for advanced manufacturing education. Integrated STEM Education Conference (ISEC), 2012 IEEE 2nd, IEEE.

Jones, B.F., Rasmussen, C.M., Moffitt, M.C., 1997. Real-Life Problem Solving: A Collaborative Approach to Interdisciplinary Learning. American Psychological Association.

Mills, J.E., Treagust, D.F., 2003. Engineering education—is problem-based or project-based learning the answer? Aust. J. Eng. Educ. 3 (2), 2–16.

Moon, Y.B., 2004. Manufacturing education at Syracuse University. Int. J. Eng. Educ. 20 (4), 578–585.

Moskowitz, H., Ward, J., 1998. A three-phase approach to instilling a continuous learning culture in manufacturing education and training. Prod. Oper. Manag. 7 (2), 201–209.

Pramanik, A.I., Islam, M.N., 2013. Technology tools and approaches to improve undergraduate education. Int. J. Res. Educ. Method. 4 (1), 390–400.

Pramanik, A., Islam, M., 2014a. Introduction of a new software package in teaching design methodology for material selection. Int. J. Inform. Educ. Technol. 4 (4), 360–363.

Pramanik, A. and M. N. Islam (2014b). "Module-based teaching of mechanical design." Using Technology Tools to Innovate Assessment, Reporting, and Teaching Practices in Engineering Education: 60.

Ram, B., Sarin, S., Park, E., Mintz, P., 1999. Providing manufacturing experiences to industrial engineering students through an extension program. Frontiers in education conference, 1999. FIE'99. 29th annual, IEEE.

Sackett, P., McCluney, D., 1992. Inter enterprise CIM—a mechanism for graduate education. Robot. Comput. Integr. Manuf. 9 (1), 9–13.

Seymour, E., 1995. Revisiting the "problem iceberg": science, mathematics, and engineering students still chilled out. J. Coll. Sci. Teach. 24, 392.

Thomas, J.W., 1999. Project Based Learning: A Handbook for Middle and High School Teachers. Buck Institute for Education.

Thomas, J.W., 2000. A Review of Research on Project-Based Learning.

Tobias, S., 1992. Revitalizing Undergraduate Science: Why some Things Work and Most Don't. An Occasional Paper on Neglected Problems in Science Education, ERIC.

Todd, R.H., Red, W.E., Magleby, S.P., Coe, S., 2001. Manufacturing: a strategic opportunity for engineering education. J. Eng. Educ. 90 (3), 397–405.

Ziemian, C., Sharma, M., 2008. Adapting learning factory concepts towards integrated manufacturing education. Int. J. Eng. Educ. 24 (1), 199–210.

CHAPTER 4

Friction Compensation in the Compression Test

P. Christiansen*, P.A.F. Martins[†], N. Bay*

*Department of Mechanical Engineering, Technical University of Denmark, Lyngby, Denmark
[†]IDMEC, Instituto Superior Tecnico, University of Lisbon, Lisbon, Portugal

1 INTRODUCTION

The compression test consists of the upsetting of a cylindrical test specimen between flat, parallel die platens in order to obtain the stress–strain curve of metallic materials. From a metal forming point of view, the stress–strain curve is one of the most important material data for modeling the mechanical behavior of metals because it is utilized to describe strain hardening, to set up the nonlinear constitutive equations of plasticity and to estimate the pressures and forces applied on workpiece and tools, among other scientific and practical engineering utilizations.

If it were possible to completely eliminate friction from the upset compression tests, the stress–strain curve $\sigma(\varepsilon)$ could easily be determined from the experimental values of force F and displacement $\Delta h = h_0 - h$ because the specimens would deform homogeneously and the diameter d would be uniformly constant along the height h (Fig. 1A).

However, there are several reasons for not achieving homogeneous plastic deformation conditions in daily practice. First, the overall experimental procedure is strongly dependent on the quality of the lubricants applied on both specimens and die platens. Second, it is impossible to ensure frictionless conditions. Even with the most efficient lubricants, residual friction appears in the upset compression test resulting in "barrelling" (Fig. 1B). This amount of residual friction needs to be identified and taken into account when calculating the stress–strain curve. And third, even in case of using specially designed test specimens such as that proposed by Rastegaev (1940), in which a lubricant reservoir is included on the top and bottom surfaces in order to encapsulate the lubricant and prevent direct metal to metal contact on most of the interface between specimen and platens, there will still be signs of friction in the sealing edge as well as in the remaining central region, when the pressurized lubricant in the reservoir escapes outward.

Manufacturing Engineering Education
https://doi.org/10.1016/B978-0-08-101247-5.00004-6

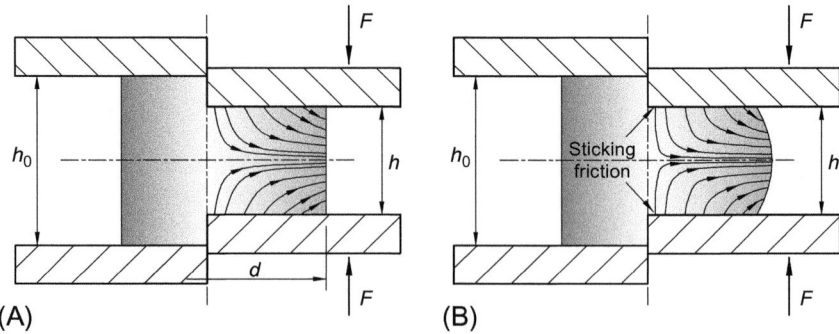

Fig. 1 Upsetting of a cylindrical test specimen under (A) homogeneous (frictionless) and (B) nonhomogeneous (friction) conditions with schematic representation of material flow (Bay and Gerved, 1987).

The alternative use of thin sheets of teflon to ensure insignificant influence of friction is furthermore limited by tearing of the sheets of teflon and subsequent metal to metal contact between specimen and tool platens. Moreover, when the aim is to perform upset compression tests in warm and hot forming regimes, it is difficult, if not impossible, to avoid a significant amount of friction even with good lubricants, and lubricants like, for instance, teflon sheets or conventional oils may be inapplicable due to the breakdown at elevated temperatures. Therefore, it is of great importance to be able to correct the stress-strain curve that is directly determined from the upset compression tests with friction.

Cook and Larke (1945) were among the first researchers to remove the frictional work done in upset compression tests. They proposed a methodology in which four specimens with different initial d_0/h_0 ratios (0.5, 1, 2, and 3) are compressed under constant lubrication conditions to preset height reductions $R = (h_0 - h)/h_0$ and their results extrapolated in a systematic way to give the desired flow curve for a specimen of zero d/h ratio (Fig. 2).

The justification behind Cook and Larke's methodology is based on the fact that as the height h of the compression test specimens increases, the influence of the sticking region between the compression platen and the workpiece (Fig. 1B) decreases so that in the limit (when the specimen has an infinite height h) this influence is negligible and plastic deformation may be considered homogeneous. The reason why the initial ratios d_0/h_0 are downwards limited to 0.5 is to prevent specimens from buckling instead of being ideally compressed in its height direction.

Alexander and Brewer (1963) revisited the utilization of Cook and Larke's method and proposed a modification based on the utilization of

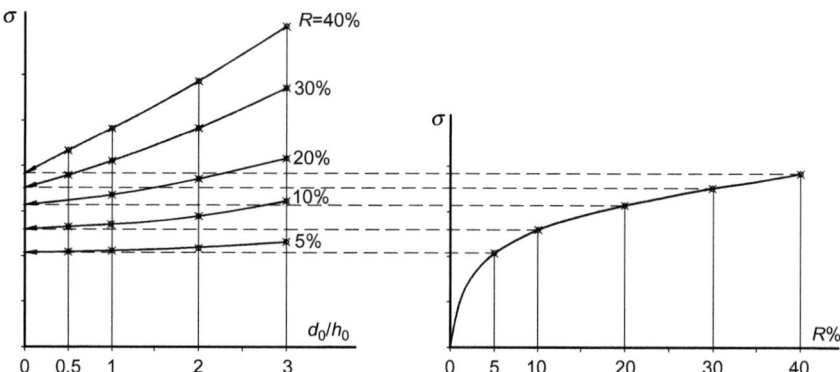

Fig. 2 Schematic representation of Cook and Larke's (1945) method for removing the frictional effects from the stress-strain curve.

equal values of force instead of equal values of reduction in height, as it was initially proposed by Cook and Larke (1945). If, in addition to this, the test is performed with increments of force instead of continuous loading, as it was originally carried out by Watts and Ford (1955) in case of the plane strain compression of strips between parallel platens, it is possible to remove the specimens from the testing machine at regular intervals to renew lubrication and significantly reduce the size of friction effects on the experimental force. Removal of the specimens from the testing machine also allows the height to be measured without having to account for the elastic deformation of the tools.

Woodward (1977) changed the extrapolating procedure associated with the different variants of Cook and Larke's method into an interpolating procedure based on the utilization of a corrective mathematical function obtained from Avitzur's (1968) upper bound solution for the compression of a cylinder between flat parallel platens. The method makes use of two upset compression tests instead of the four upset compression tests required by Cook and Larke's method, but requires the specimens and the platens to be very well-lubricated. In fact, Woodward (1977) suggested the teflon sheets to be renewed at regular intervals during compression so that uniform compression is achieved and the effect of friction on the measured compression load is therefore initially small.

Osakada et al. (1981) obtained stress–strain curves of metals by means of an inverse analysis using a rigid–perfectly plastic finite element method. Various compensations due to strain hardening and temperature increase had to be performed and eventually comparisons with upsetting tests using thin teflon sheets were performed. The method of applying finite element analysis may be limited by the availability of a finite element computer program in the laboratory or workshop.

Han (2002) modified Woodward's method (1977) by assuming the flow stress for a given material to be independent from the geometry of the specimen, and thus, was able to set up an objective function dependent on friction. The coefficient of friction and the flow stress are determined by means of an inverse computational procedure that makes use of the objective function built upon the slab method solution for the compression of a cylinder between flat parallel die platens (Mielnik, 1991).

Xinbo et al. (2002) proposed an alternative procedure in which an objective function and subsequent optimization procedure are built upon the differences between the experimental and finite element estimates of the compression force.

In addition to the above-mentioned procedures for determining the stress-strain curve directly from the force-displacement evolution obtained in the experiments and indirectly from the utilization of mathematical or numerical procedures that eliminate the effects of friction from the compression force, there are other approaches focused on the characterization of the friction directly from the barreled surface of the upset compression test specimens. Tan et al. (1999), for example, developed a procedure to determine friction by applying the relative shrinking ratio of the original end surface of the test specimens as a calibration parameter. Once friction is determined, it is easy to determine the stress–strain curve of the material from the experimental measurements of force and displacement.

From what was mentioned above, it may be concluded that the problem of determining the stress-strain curve directly from upset compression tests with friction requires specific methods and procedures that are different from those included in classical tribology publications dealing with friction, lubrication, and surface/interface kinematics (Wilson, 1979).

Under these circumstances, this chapter describes a simple and effective methodology to determine the stress-strain curve from the experimental measurements of force and displacement in testing conditions where friction between the cylindrical specimen and the platens would cause barreling of the outer surface.

The methodology was proposed by Christiansen et al. (2016) and, in contrast to other optimization-based methods, it accounts for the changes in friction arising from the differences in pressure at the center and at the edges of the test specimens by making use of objective functions that are built upon the slab method of analysis using three different friction models based on Coulomb, constant, and combined friction models. Combined friction models (not to be confused with mixed lubrication) make use of Coulomb friction for modeling the low pressures found at the edge of the specimens and constant friction for modeling the high pressures found at the center of the specimens.

The computer implementation of the above-mentioned methodology is comprehensively described and the computer program, written in MATLAB that was utilized for determining the stress-strain curves included in this chapter, is provided for the readers interested in determining the stress-strain curve of metallic materials in daily practice.

The overall methodology and computer implementation are assessed by means of upset compression tests performed with Rastegaev (1940) test specimens and cylindrical test specimens lubricated with grease or thin sheets of teflon on top and bottom surfaces. Friction was determined independently by means of ring compression tests and the corresponding values were utilized to validate the predicted values of friction that were determined from the new proposed methodology. The resulting stress-strain curves after friction compensation are fitted by a well-known strain hardening material model.

2 THEORETICAL BACKGROUND

2.1 Objective Function

The proposed methodology provides a mathematical approximation of the stress-strain curve $\sigma(\varepsilon)$ by means of the strain hardening material models that are listed in Table 1 from the experimental evolution of the force with displacement in upset compression tests performed with friction. The methodology is developed for cold forming, where rate effects are negligible and the effective stress or flow stress $\bar{\sigma}$ is commonly assumed to be a function only of the effective plastic strain $\bar{\varepsilon}$ (hereafter designated as "the effective strain"). Strain rate and/or temperature effects could also be included in the overall approach, but are not considered in the present chapter.

In Table 1, K is a constant for each strain hardening model, σ_Y is the yield stress (true stress at the onset of plastic deformation), $\bar{\varepsilon}_0$ is a prestrain, σ_{sat} is the saturation stress, and ε_c is a constant that determines the rate at which the stress from its initial value tends to reach the saturation stress σ_{sat}.

Table 1 Strain hardening material models

Ludwik (1909)	$\bar{\sigma} = K\bar{\varepsilon}^n + \sigma_Y$
Hollomon (1945)	$\bar{\sigma} = K\bar{\varepsilon}^n$
Swift (1952)	$\bar{\sigma} = K(\bar{\varepsilon}_0 + \bar{\varepsilon})^n$
Voce (1955)	$\bar{\sigma} = \sigma_{sat} - (\sigma_{sat} - \sigma_Y)\left\{1 - \exp\left(-\dfrac{\bar{\varepsilon}}{\varepsilon_c}\right)\right\}$

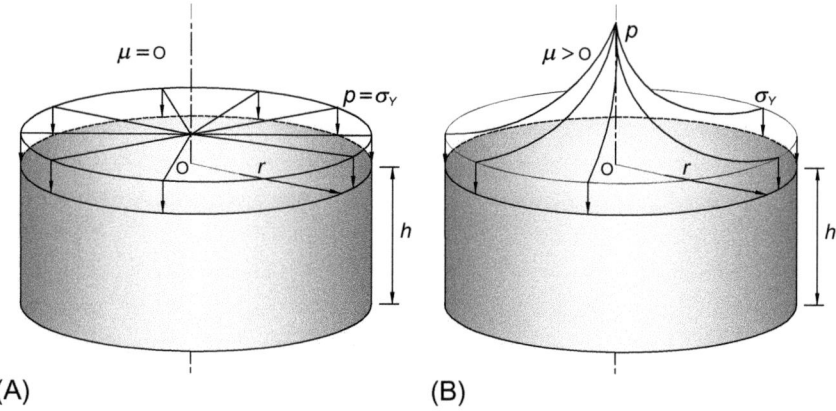

Fig. 3 Schematic representation of the pointwise distribution of pressure obtained from the application of the slab method for the upset compression of a cylindrical test specimen (A) without and (B) with friction.

The effective strain under frictionless compression conditions is given by $\bar{\varepsilon} = \ln(h_0/h_1)$, where h_0 is the initial and h_1 is the final height of the cylinder specimens. The corresponding effective stress $\bar{\sigma}$ is computed by,

$$\bar{\sigma} = p_{avg} \tag{1}$$

where p_{avg} is the average surface pressure that coincides with the pointwise distribution of pressure p in the absence of friction (Fig. 3).

The objective function for determining the stress–strain curve is given by $f(M, \tau_s) = |p_{avg}^{exp} - p_{avg}|$, where M denotes the independent parameters of the strain hardening material model, τ_s is frictional shear stress at the contact interface between the specimen and the platens, p_{avg}^{exp} is the experimental average surface pressure, and p_{avg} is the corresponding estimate of average pressure obtained from the slab method solution for the compression of a cylindrical test specimen with diameter $d = 2r$ and height h between flat parallel platens (Han, 2002). Two models are taken into consideration for modeling the frictional effects at the contact interface between the specimen and the platens: the Coulomb friction model given by $\tau_s = \mu p$, where μ is the friction coefficient and p is the normal pressure,

$$p_{avg} = 2\bar{\sigma} \left(\frac{h}{\mu d}\right)^2 \left[\exp\left(\frac{\mu d}{h}\right) - \frac{\mu d}{h} - 1\right] \tag{2}$$

and the constant friction model $\tau_s = mk$, where m is the friction factor $(0 \le m \le 1)$ and k is the yield stress of the material in pure shear,

$$p_{avg} = \bar{\sigma}\left(1 + \frac{md}{3\sqrt{3}h}\right) \qquad (3)$$

Similarly to the slab method of analysis, the mathematical approximation of the stress–strain curve by means of the new proposed procedure is based on the following three assumptions; (i) the principal axes are in the directions of the applied loads, (ii) the effects of friction do not change the directions of the principal axes, and (iii) plane sections remain plane during compression.

In contrast to other approaches available in literature, the proposed methodology is capable of combining the two above estimates of average pressure (Eqs. (2) and (3)) in order to account for the differences in lubrication that are found between the center and the edge of the cylindrical test specimens. These differences are due to the fact that during compression, lubricant runs out of the edges and the barreled surface folds up onto the compression platens giving rise to dry metal-to-metal contact. As a result of this, the frictional shear stresses τ_s are higher at the edges than in the central region of the specimens where lubricant becomes trapped. The same situation applies with the utilization of thin sheets of teflon, which are cut out by the edges of the specimens during compression.

In case of combining the two above-mentioned friction models, it is considered that the constant friction model is the most adequate for the central region of the specimen and the Coulomb friction model for the outer region $(0 \le r_g \le r)$, where r_g is the transition radius between the two friction models:

$$p_{avg} = \bar{\sigma}\left(\frac{m}{\sqrt{3}}\left[\frac{1}{\mu}\left(\frac{r_g}{r}\right)^2 + \frac{2r_g^3}{3hr^2}\right] + \frac{1}{2\mu^2}\left(\frac{h}{r}\right)^2\left[\left(1 + \frac{2\mu r_g}{h}\right)\exp\left\{\frac{2\mu\left(r - r_g\right)}{h}\right\} - \frac{2\mu r}{h} - 1\right]\right) \qquad (4)$$

In case $r_g = 0$, only the Coulomb friction model is utilized and, in case of $r_g = r$ and $\mu = m/\sqrt{3}$, only the constant friction model is utilized. For numerical implementation purposes, combination of the two friction models required the utilization of a modified version of Eq. (4), in which the radius

r_g is expressed by means of a fixed fraction x of the outer radius r ($r_g = xr$ with $0 \leq x \leq 1$).

$$p_{avg} = \bar{\sigma} \left(\frac{m}{\sqrt{3}} \left[\frac{1}{\mu} x^2 + \frac{2x^3 r}{3h} \right] + \frac{1}{2\mu^2} \left(\frac{h}{r} \right)^2 \left[\left(1 + \frac{2\mu x r}{h} \right) \exp \left\{ \frac{2\mu(1-x)r}{h} \right\} - \frac{2\mu r}{h} - 1 \right] \right)$$

(5)

Eq. (5) will be used later for the comparison between experimental and computed compression forces.

2.2 Computer Implementation

The proposed methodology for determining the stress-strain curve starts with a first guess of the stress-strain curve obtained from the experimental evolution of the force with displacement retrieved from the compression test specimens. The first guess assumes that the compression test specimens experience homogenous plastic deformation (Fig. 3A).

The stress-strain curve of the material is fitted by one of the strain hardening material models listed in Table 1 and the corresponding independent parameters (e.g., K and n in case of the Hollomon $\bar{\sigma} = K\bar{\varepsilon}^n$ strain hardening material model) are subsequently determined by means of a numerical procedure aimed at minimizing the objective function $f(M, \tau_s, x) = |p_{avg}^{exp} - p_{avg}|$. The minimum is considered to be reached when the change in the solution becomes sufficiently small $\Delta |p_{avg}^{exp} - p_{avg}| / |p_{avg}^{exp}| < \delta$, where $\Delta |p_{avg}^{exp} - p_{avg}|$ is the difference between the previously best and current best residuals.

The simplest algorithm that can be utilized to determine the independent parameters of the strain hardening material model and of the friction coefficient (and/or friction factor) brackets the root of $f(M, \tau_s, x)$ within a search interval $[M_a, \tau_{sa}, x_a; M_b, \tau_{sb}, x_b]$.

Knowledge of the typical order of magnitude of the independent parameters and friction values helps limiting the search interval to values in ranges that are meaningful from a plastic deformation point of view (e.g., $0 \leq n \leq 0.5$, $0 \leq \mu \leq 1/\sqrt{3}$, and $0 \leq m \leq 1$).

The computer program, written in MATLAB, which is provided in what follows, contains the above-mentioned numerical procedure and is structured into three main parts: (i) input of experimental data, (ii) minimization of the difference between computed and measured average pressures, and (iii) output of the stress-strain curve.

2.2.1 Input Experimental Data

The experimental data is supplied in a file containing the following information:

```
r                % Radius of specimen during upsetting (vector)
h                % Height of specimen during upsetting (vector)
Epsilon          % Experimental effective strain (vector)
p_exp            % Experimental average surface pressure (vector)
```

The source code is the following:

```
K_min                                           % Minimum K

K_max                                           % Maximum K

n_min                                           % Minimum n

n_max                                           % Maximum n

Friction_my_min                                 % Minimum friction coefficient

Friction_my_max                                 % Maximum friction coefficient

Friction_m_min                                  % Minimum friction factor

Friction_m_max                                  % Maximum friction factor

Ratio_min                                       % Minimum friction ratio

Ratio_max                                       % Maximum friction ratio

delta_K = K_max - K_min                         % Search interval of K

delta_n = n_max - n_min                         % Search interval of n

delta_my = Friction_my_max - Friction_my_min    % Search interval of Coulomb friction coef.

delta_m = Friction_m_max - Friction_m_min       % Search interval of Tresca friction factor

delta_x = Ratio_max - Ratio_min                 % Search interval x

i_max                                           % Maximum number of iterations

Residual_best = 1E99                            % Initialization

tol                                             % Convergence tolerance
```

2.2.2 Minimization of the Difference Between Experimental and Computed Average Pressures

This part of the computer program is based on an iterative cycle that starts by calculating the average surface pressure p_{avg}, follows by determining the difference $|p_{avg}^{exp} - p_{avg}|$ between computed and measured average pressures, and ends by checking the norm $\Delta|p_{avg}^{exp} - p_{avg}|/|p_{avg}^{exp}| < \delta$. The source code is the following:

```
for i=1:i_max % Loop from 1 to maximum number of iterations

    K = K_min+rand(1)*delta_K
    n = n_min+rand(1)*delta_n
```

```
my = my_min+rand(1)*delta_my
m = m_min+rand(1)*delta_m
x = x_min+rand(1)*delta_x
% Computation of average surface pressure
p = K*Epsilon.^n.*(m/sqrt(3).*(1/my*x^2+2*x^3*r/3./h)
    +1/2/my^2*(h./r).^2.*((1+2*my*x*r./h).*exp(2*my*r.*
    (1-x)./h)-2*my*r./h-1) );
Residual = norm(p-p_exp) % Computation of difference between
                           computed and measured average pressure

    if Residual < Residual_best % Control if new set of constants is better
                             than previous optimum
        K_best = K      % Optimum strength coefficient
        n_best = n      % Optimum strain hardening exponent
        my_best = my    % Optimum Coulomb friction coefficient
        m_best = m      % Optimum friction factor
        x_best = x      % Optimum x
        delta = (Residual_best-Residual)/Sigma_norm;
        Residual_best = Residual
        if delta < tol % Control for convergence
            break
        end
    end
end
end
```

2.2.3 Output of the Stress-Strain Curve

The output of the computer program provides the constants of the different strain hardening models listed in Table 1.

3 EXPERIMENTATION

3.1 Upsetting of Cylindrical and Rastegaev Test Specimens

The experiments for investigating the accuracy and reliability of the proposed methodology were performed on Aluminum Al2S (99.7% Al, 0.2% Fe, 0.1% Si) supplied in the form of 30 mm diameter rods. The cylindrical and Rastegaev test specimens were machined in accordance to the geometries shown in Fig. 4 and the experiments were carried out with the material in the "as-supplied" condition.

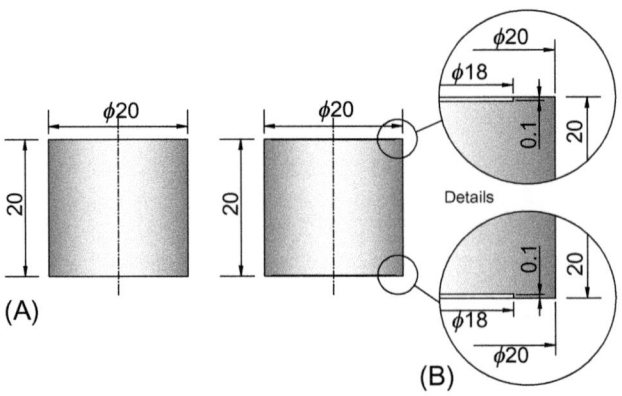

Fig. 4 Geometries of the (A) cylindrical and (B) Rastegaev compression test specimens.

The upset compression tests were performed in a hydraulic press with constant moving cross-head speed and the force-displacement evolution was recorded on a personal computer.

The compression die platens were cleaned with ethanol before each experiment. The cylinder test specimens were lubricated with Molykote DX paste or teflon sheets on the top and bottom surfaces before compression in order to reduce friction. These two different types of compression tests are denoted hereafter as "Cylinder" and "Teflon," respectively. The Rastegaev test specimens were lubricated with Molykote DX paste on the top and bottom ends, including the grooves, before compression. This experiment is denoted hereafter as "Rastegaev." Examples of the compression test specimens, before and after compression, and of the teflon sheets are shown in Fig. 5.

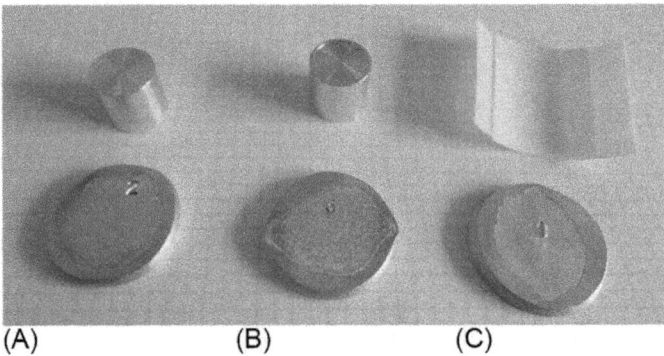

Fig. 5 (A) Cylinder, (B) Rastegaev, and (C) Teflon compression test specimens before and after deformation.

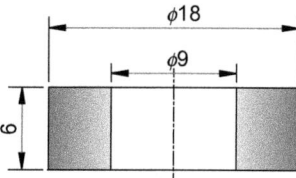

Fig. 6 Geometry of the ring test specimens.

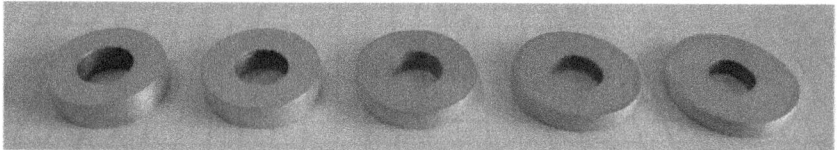

Fig. 7 Rings test specimens after upsetting with Molykote DX paste at increasing reduction (*from left to right*).

3.2 Upsetting of Ring Test Specimens

The ring test specimens were machined from the same aluminum rod as the compression test specimens. The dimensions are shown in Fig. 6.

The rings were also lubricated with Molykote DX paste or with sheets of teflon. Before upsetting, the inner and outer diameter and the height were measured using a Vernier caliper. After each upset increment, the smallest inner diameter and final height were also measured. Fig. 7 shows the geometry of the rings at increasing reductions.

The compression die platens used in both compression and ring compression tests had an average roughness $R_a = 0.2$ µm. However, it must be emphasized that surface roughness has no direct relevance in the proposed methodology to determine the stress–strain curve of metallic materials. In fact, the friction compensation procedure developed by the authors is successful in eliminating all the parameters that are responsible for deviating the evolution of the force with displacement obtained in the real upsetting of cylindrical test specimens from that obtained under homogeneous, frictionless, upsetting of cylindrical test specimens, during which lubrication and surface finish of both specimens and compression die platens are supposed to be ideal/perfect.

4 RESULTS AND DISCUSSION

4.1 Calibration of Friction

The results of the ring test experiments are disclosed in Fig. 8. The calibration curves for both Coulomb and constant friction models were obtained

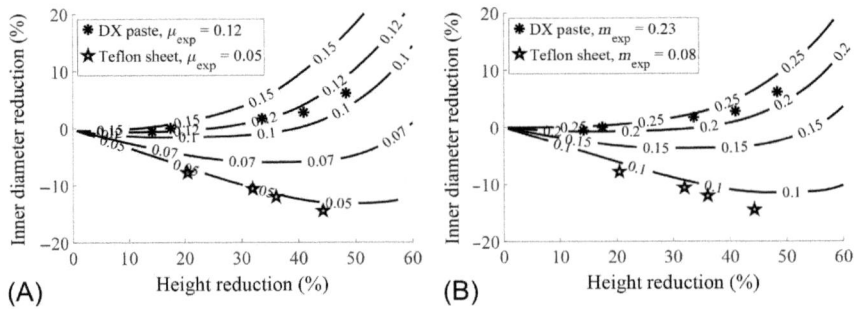

Fig. 8 Friction calibration curves and experimental results for the ring compression test. (A) Coulomb friction model—μ and (B) Constant friction model—m.

with the in-house finite element computer program I-form (Nielsen et al., 2013) using a stress–strain curve $\bar{\sigma} = 131\bar{\varepsilon}^{0.26}$ MPa (refer to Fig. 8A and B).

The average values of the friction coefficient $\mu_{\exp}$ and of the friction factor $m_{\exp}$ were determined from linear interpolation between the lines. As seen, the average Coulomb friction coefficient is $\mu_{\exp} = 0.12$ and the average friction factor is $m_{\exp} = 0.23$ when applying Molykote DX paste. The standard deviation is 0.0087 and 0.0012, respectively. For the teflon sheets, the average Coulomb friction coefficient is $\mu_{\exp} = 0.05$ and the average friction factor $m_{\exp} = 0.08$. The standard deviation is 0.0109 and 0.0075, respectively.

4.2 Stress-Strain Curves

The determination of the different constants of the strain hardening material models listed in Table 1 from the experimental evolution of the force with displacement obtained in Cylindrical and Rastegaev compression test specimens followed the methodology that was previously described in Section 2. For presentation purposes, it was decided to use the Hollomon model and, for readability purposes, it was also decided to include the results obtained from the first stage of the numerical procedure. However, the methodology can be easily applied to the other strain hardening material models that are listed in Table 1.

4.2.1 Hollomon—Frictionless

The first stage of the numerical procedure assumes the results of the compression tests to be obtained under homogenous (frictionless) plastic deformation. This allows determining an initial guess for the independent parameters K and n (strength coefficient and strain hardening exponent)

Table 2 Best fit constants based on Hollomon fit

Experiment	Strength coefficient K (MPa)	Strain hardening exponent n	R^2
Cylinder	156	0.38	0.9729
Rastegaev	153	0.36	0.9774
Teflon	120	0.28	0.9967

of the material strain hardening model (Table 2). The coefficient of determination (R^2) is listed in Table 2 for each compression test.

As seen in Table 2, there are differences in the results obtained from the compression tests performed with teflon sheets and those performed with Molykote DX paste. Differences are also found in the results obtained from the Rastegaev type specimen. The cylinder with Molykote DX paste and the Rastegaev test specimens give approximately similar values of K and n, but these are significantly larger than those obtained for the cylinder with teflon sheets.

Fig. 9 shows the measured and computed evolutions of the average surface pressure with the effective strain, using the material parameters that are included in Table 2. As seen, both the Cylinder and the Rastegaev test specimens deviate from the Hollomon stress-strain curve. In contrast, the specimen with teflon sheets presents a fair agreement with the assumed frictionless upsetting of a Hollomon strain hardening material.

4.2.2 Hollomon—Coulomb Friction
The second stage of the proposed numerical procedure takes friction into account and determines the independent parameters K and n of the material

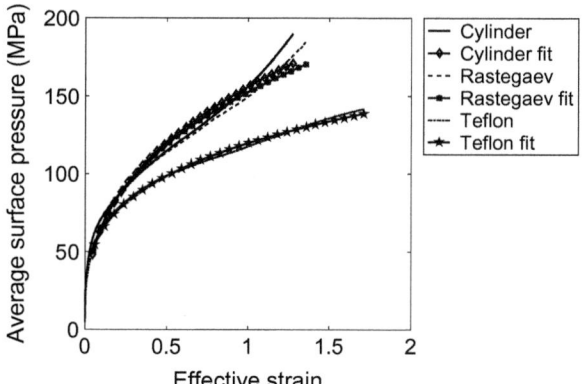

Fig. 9 Compression tests with Hollomon fit.

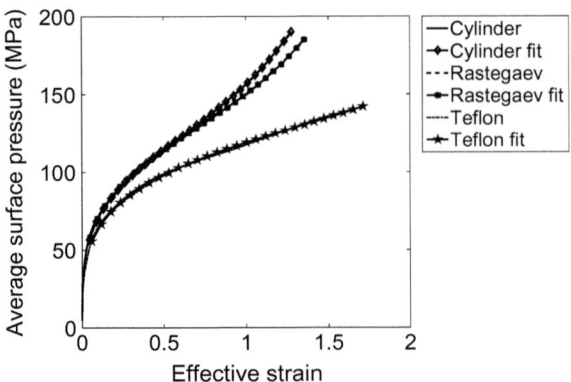

Fig. 10 Compression tests with Hollomon-Coulomb friction fit.

strain hardening model simultaneously with the coefficient of friction μ (in case of assuming a Coulomb friction model, Eq. (2)).

Fig. 10 shows the evolution of the average surface pressure with the effective strain and the constants of the material strain hardening model are given in Table 3. The convergence criterion δ was set equal to 10^{-4} and the overall CPU time was below 10 s in a standard laptop computer.

As can be seen in Fig. 10, there is a good agreement between the final results provided by the proposed models and the experiments. Table 3 also shows that the independent parameters K and n of the material strain hardening model of the Cylinder and Rastegaev experiments are closer to the Teflon test case than in the first guess, when frictionless conditions were assumed.

Results also show that the friction coefficient of the Cylinder test ($\mu = 0.15$) is in good agreement with the friction coefficient obtained from the ring tests ($\mu_{exp} = 0.12$). The Rastegaev test specimen shows some level of residual friction ($\mu = 0.11$) in reasonable agreement with the ring test.

The friction coefficient of the compression test performed with teflon sheets ($\mu = 0.01$) is smaller than that found in the ring test. This may be

Table 3 Best fit constants based on Hollomon-Coulomb friction fit

Experiment	Strength coefficient K (MPa)	Strain hardening exponent n	Friction coefficient μ	R^2
Cylinder	124	0.27	0.15	0.9997
Rastegaev	126	0.27	0.11	0.9997
Teflon	117	0.26	0.01	0.9979

due to the rings cutting into the teflon sheets, thereby increasing friction in the ring test as compared to the compression test.

4.2.3 Hollomon—Constant Friction
In case of the constant friction model, the numerical procedure takes into account Eq. (3) and the corresponding values of K and n are computed together with the friction factor m. Fig. 11 shows the evolution of the average surface pressure with the effective strain and Table 4 shows the calculated values of the constants of the strain hardening material model.

Fig. 11 shows a good agreement between Eq. (3) and the experiments. Table 4 shows a reasonable agreement between the three strength coefficients and the strain hardening exponents. However, the friction factor for the Cylinder experiment ($m=0.38$) is too large when compared with that obtained from ring tests ($m_{exp}=0.23$). As a result of this, it may be concluded that frictional conditions in the Cylinder are better modeled by the Coulomb friction model for the experimental conditions that were utilized in this investigation.

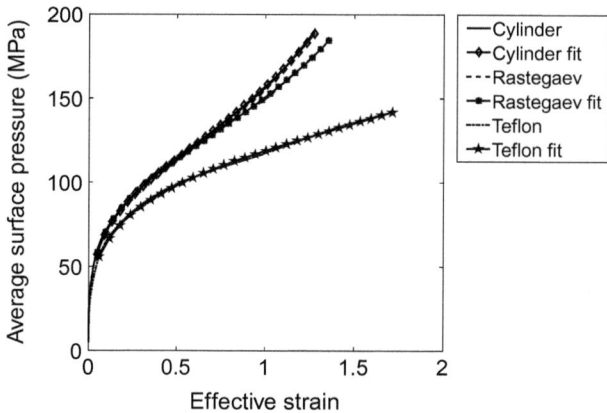

Fig. 11 Compression tests with Hollomon-Constant friction fit.

Table 4 Best fit constants based on Hollomon-Constant friction fit

Experiment	Strength coefficient K (MPa)	Strain hardening exponent n	Friction factor m	R^2
Cylinder	118	0.26	0.38	0.9996
Rastegaev	123	0.26	0.27	0.9997
Teflon	117	0.26	0.02	0.9979

In contrast, the Rastegaev experiment shows a friction factor similar to the one obtained in the ring tests. This result is easy to understand because the contact geometry of both tests (along a ring zone) is somewhat identical.

The Teflon experiment presents once again a friction factor that is very close to frictionless conditions and smaller than that found in the ring test.

4.2.4 Hollomon—Combined Friction

In case of the combined friction model, the numerical procedure makes use of Eq. (5). The K and n values are computed together with the friction coefficient μ, the friction factor m, and the radius ratio $x = r_g/r$, according to the objective function $f(M, \tau_s, x)$. Fig. 12 provides the evolution of the average surface pressure with the effective strain and the material strain hardening independent parameters are given in Table 5.

Results show a reasonable agreement between Eq. (5) and the experiments. Table 5 indicates a fair agreement between the strength coefficient K and the strain hardening exponent n of the three experiments.

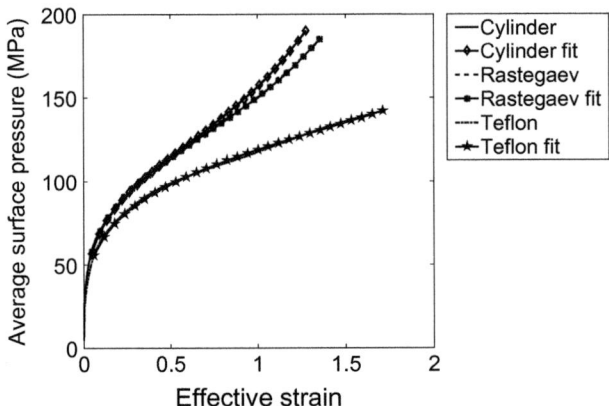

Fig. 12 Compression tests with Hollomon-combined friction fit.

Table 5 Best fit constants based on Hollomon-combined friction fit

Experiment	Strength coefficient K (MPa)	Strain hardening exponent n	Friction factor m	Friction coefficient μ	Radius ratio x	R^2
Cylinder	124	0.27	0.21	0.15	0.10	0.9997
Rastegaev	127	0.27	0.10	0.11	0.08	0.9997
Teflon	117	0.26	0.02	0.01	0.00	0.9979

In case of the Cylinder, there is also a good agreement between the calculated and experimental values of the friction factor and the friction coefficient ($\mu = 0.15$ and $m = 0.21$ vs $\mu_{exp} = 0.12$ and $m_{exp} = 0.23$). It can also be concluded that, although the radius ratio x is fairly small for all the experiments, it is large enough to justify the need for using a combined friction model in case of the Cylinder and the Rastegaev specimens. This last result explains the reason why the Cylinder experiment predicts a somewhat larger value of the friction factor, when applying Eq. (3), than that found by means of the ring test experiment.

5 CONCLUSIONS

An innovative experimental and numerical methodology has been proposed to compensate for friction when performing upsetting compression tests. The method has been tested on two different test specimens; Cylinder and Rastegaev. The cylinders were lubricated with either grease or teflon sheets on top and bottom surfaces, while the Rastegaev's specimens were lubricated with grease.

When the stress-strain curve is directly obtained from the experimentally measured force-displacement curves, there are differences resulting from the type of specimen and lubrication procedure. However, these differences disappear when the effect of friction on the compression force is corrected by means of the proposed methodology and similar values of strength coefficient and of the strain hardening exponent are obtained.

The application of a combined friction model making use of Coulomb friction for modeling the low pressures found at the edge of the specimens and constant friction for modeling the high pressures found at the center of the specimens is considered to provide the best estimate of the stress-strain curve because the friction coefficient and the friction factor of the upset compression tests are similar to those obtained by means of independent ring compression tests.

ACKNOWLEDGMENTS

The authors would like to acknowledge the support provided by The Danish Council for Independent Research under the grant number DFF-4005-00130. Paulo Martins would also like to acknowledge the support provided by Fundação para a Ciência e a Tecnologia of Portugal under LAETA-UID/EMS/50022/2013 and PDTC/EMS-TEC/0626/2014.

REFERENCES

Alexander, J.M., Brewer, R.C., 1963. Manufacturing Properties of Materials. Van Nostrand, London.

Avitzur, B., 1968. Metal Forming: Processes and Analysis. McGraw Hill, New York.

Bay, N., Gerved, G., 1987. Tool/workpiece interface stresses in simple upsetting. J. Mech. Work. Technol. 14, 263–282.

Christiansen, P., Martins, P.A.F., Bay, N., 2016. Friction compensation in the upsetting of cylindrical test specimens. Exp. Mech. 56, 1271–1279.

Cook, M., Larke, E.C., 1945. Resistance of copper and copper alloys to homogeneous deformation in compression. J. Inst. Metals 71, 371–390.

Han, H., 2002. The validity of the mathematical models evaluated by two-specimen method under the unknown coefficient of friction and flow stress. J. Mater. Process. Technol. 122, 386–396.

Hollomon, J.H., 1945. Tensile deformation. T. Am. I. Min. Met. Eng 162, 268–290.

Ludwik, P., 1909. Elemente der Technologischen Mechanik. Springer Verlag, Berlin.

Mielnik, E., 1991. Metalworking Science and Engineering. McGraw Hill, New York.

Nielsen, C.V., Zhang, W., Alves, L.M., Bay, N., Martins, P.A.F., 2013. Modelling of Thermo-Electro-Mechanical Manufacturing Processes With Applications in Metal Forming and Resistance Welding. Springer-Verlag, London.

Osakada, K., Kawasaki, T., Mori, K., 1981. A method of determining flow stress under forming conditions. Ann. CIRP 30, 135–138.

Rastegaev, M.V., 1940. Neue methode der homogenen Stauchung. Z. Lab. 3, 354–355.

Swift, H.W., 1952. Plastic instability under plane strain. J. Mech. Phys. Solids 1, 1–18.

Tan, X., Zhang, W., Bay, N., 1999. A new friction test using simple upsetting and flow analysis. Adv. Technol. Plast. 1, 365–370.

Voce, E., 1955. A practical strain hardening function. Metallurgia 51, 219–226.

Watts, A.B., Ford, H., 1955. On the basic yield stress curve for a metal. P Inst. Mech. Eng. 169, 1141–1156.

Wilson, W.R.D., 1979. Friction and lubrication in bulk metal-forming processes. J. Appl. Metalworking 1, 1–19.

Woodward, R.L., 1977. A note on the determination of accurate flow properties from simple compression tests. Metall. Trans. A. 8A, 1833–1834.

Xinbo, L., Fubao, Z., Zhiliang, Z., 2002. Determination of metal material flow stress by the method of C-FEM. J. Mater. Process. Technol. 120, 144–150.

FURTHER READING

Chapra, S., Canale, R., 2009. Numerical Methods for Engineers. McGraw-Hill, New York.

CHAPTER 5

Appreciation of CNC Technology Through Machine Tool Upgrading by an Open Controller

George-Christopher Vosniakos, Nikolaos Zourtsanos, Nikolaos Kontogiannis

National Technical University of Athens, School of Mechanical Engineering, Section of Manufacturing Technology, Athens, Greece

1 INTRODUCTION

Older machine tools are often outdated in terms of their functionality and operator friendliness, but otherwise they are mechanically sound with or without refurbishment or replacement of some of their components (Uhlmann et al., 2017). Thus, the much discussed notion of remanufacturing may, in the case of machine tools, often come down to upgrading their controllers (Du et al., 2012; Wang et al., 2014). In the case of conventional controllers, such upgrading is often impossible due to their vendor-specific nature and uncommon interfaces, but, even if it were possible, there would be no added benefit in retaining old functionality (Park et al., 2006). Furthermore, there is no better way to consolidate the basic engineering knowledge in computer numerically controlled machine tools than upgrading a CNC machine tool, since building it from scratch is much more cumbersome.

In terms of controller architectures, mixed software–hardware have been studied in the literature. In Fei et al. (2011), an embedded system is proposed using the ARM (Advanced RISC Machine) processor and an FPGA (Field Programmable Gate Array). Communication with the host PC was achieved with a serial bus so that higher level data could be processed, concerning part program development and simulation and shop floor data management. In Wang et al. (2011), a CNC system was reconfigured by defining a new hardware architecture based on ARM, FPGA, and DSP. As for the software system reconfiguration, a development platform was proposed with a human–machine interface to manipulate CNC function library. Along similar lines, in Xu and Chen (2012) the design of a CNC system embedded

Manufacturing Engineering Education
https://doi.org/10.1016/B978-0-08-101247-5.00005-8

on ARM was discussed using a DSP motion control chip and an open-source operating system; a detailed breakdown of the system modules was provided and a good interrupt response was claimed. On the low-cost side, in Khanna et al. (2013) an open-source CNC system was presented capable of 6-axis simultaneous interpolation by embedding the necessary features in an Arduino system which processed the NISTSAI canonical code (NCC). Similarly, in Xu et al. (2012) an open-source CNC system was discussed based on a PC and a motion controller card and following a hierarchical modular structure. As an alternative in terms of distributed systems, Zhang et al. (2007) designed a hierarchical real-time network based on Ethernet/Internet, whose field-level communications met the requirements of hard real-time tasks; switched Ethernet for communication between the NC server; and the NC core computers also met the requirement of soft real-time tasks.

Several open-control architectures emerged in the previous two decades, e.g. OSEC, OSACA, JOP, and OMAC with a modular structure and standardized communication for transparent data exchange and plug-and-play adaptation (Pritschow et al., 2001). A comprehensive and systematic attempt toward an open modular architecture controller (OMAC) was presented in Ma et al. (2007) based on Windows OS, data flow representation by finite-state machines and reusable function module libraries; a 3-axis milling machine tool test bed was provided, resulting in lower development and maintenance time for the controller. These early attempts suffered from complex applications programming interface, thus they were difficult to adopt.

The software-based CNC is flexible, and lately, fast, and robust enough to support open controllers. In order to ensure adequate robustness and time response, real-time OS, middleware kernels, and smart modularization are exploited. In Martinov et al. (2015), a multiprotocol CNC system kernel was proposed addressing CANbus, as a reliable and low-cost solution for medium-priced machines. Similarly, Park et al. (2006) proposed a kernel software acting as middleware between the various CNC control function software modules and the CNC machine by using process and resource models upon which interface specifications were created. In Ji et al. (2008), a four-layer soft CNC system was introduced consisting of a GUI, non-real-time layer, driver layer, and RT-Linux as its kernel, adopting a rational design of data buffering and high precision period of the real-time threading to safeguard real-time performance.

Due to its inherent qualities and openness, Linux was used as the OS for open controllers. For instance, Zhang et al. (2003) develops an integrated environment for building an open CNC system with off-the-shelf PC hardware, open-source Linux/RT-Linux as the system platform and a Universal Serial Bus (USB) to communicate between the CNC system and machine tools. Most influentially, NIST launched EMC (LinuxCNChttp, 2016), an open-source software-oriented numerical controller running under Linux, originally using a common real-time function library, to assure real-time performance and homogeneous communications, and later using RT-Linux as Linux's real-time extension; some communication pitfalls were noted consisting of a shared memory being used as data buffer between OS processes and real-time threads, and also for communication between real-time modules (LinuxCNChttp, 2016).

Several applications of LinuxCNC (EMC2) have been reported so far. It has been used as the main controller that was interfaced with an HMI manipulating a business and production knowledge-based system in an effort to design a smart machine, i.e., one that knows its capabilities (Hentz et al., 2013). In Staroveski et al. (2011), the control of a 3-axis desktop milling machine was implemented with an aim to test most features of the software. Further details of an analogous implementation concerning upgrading of an industrial caliber machine are given in Staroveški et al. (2009). In Gutierrez and Álvares (2013), a STEP-NC adapter was developed that replaced the RS247 interface of LinuxCNC and generated the G-code for a 3-axis experimental router.

In this chapter, upgrading of a desktop CNC lathe around an open CNC controller is reported, focusing on the interfacing procedures, software, and hardware to match the functionality offered by the controller. The machine tool as such is, in fact, one of the most commonly used in research and light industrial applications. The contribution of this work lies in the detailed treatise of technical solutions making up these interfaces in the light of embedding CNC of machine tools into a manufacturing education curriculum. The original specification of the machine is presented in Section 2, the new hardware interfaces in Section 3, and the open controller software interfaces in Section 4. Section 5 presents the results of turning a sample aluminum part and summarizes pertinent conclusions.

2 ORIGINAL SPECIFICATION OF THE MACHINE TOOL

The machine tool that is upgraded in this work is an EMCO COMPACT 5 CNC lathe with two axes and a footprint of $830 \times 620\,\text{mm}^2$ (see Fig. 1). This is one of the most popular bench-top machine tools that has typically been

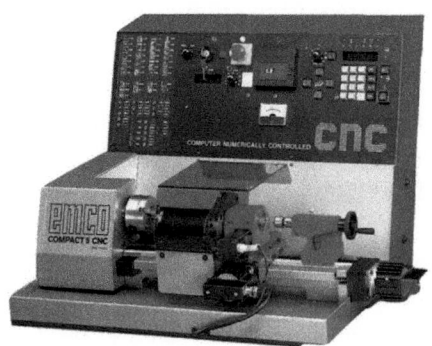

Fig. 1 Emco Compact 5 CNC lathe: (A) initial and (B) final.

used in education and light production environments, in the last decade, because it has a solid mechanical construction and possesses reliable electrical and electronics components, most of which are to be retained during upgrading of the machine.

2.1 Structure

The main spindle has a diameter of 16 mm with an MT-2 Morse cone at its end. It is driven by an 80 V dc permanent magnet motor with an input power of 500 W/300 W (see Fig. 2A). The speed range is changed in six steps between 50 and 3200 rpm using a belt and pulley system (see Fig. 2B). The desired speed was originally set by a potentiometer. Spindle speed is measured through a photocell and a series of peripheral holes on a disk rotating together with the spindle (see Fig. 2B). Similarly, the spindle position is sensed by a slot and photocell system, which is typically necessary for thread cutting.

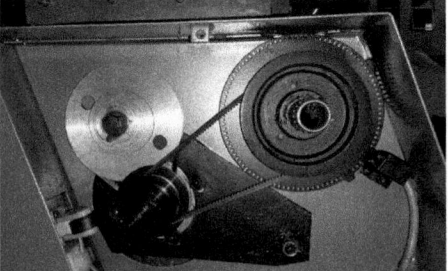

Fig. 2 Main spindle: (A) motor and (B) power transmission.

Axes X and Z are driven by unipolar stepping motors with a five-step and 0.5 Nm torque. These transmit motion via a belt to the ball screws, each of which has a pitch of 2.5 mm (see Fig. 3) yielding a linear resolution of 13.89 μm per step of the motor. The digital indicator on the console (see Fig. 1) has a resolution of 0.01 mm.

Rapid motion speed can reach 700 mm/min for each axis, while the cutting speed can range between 2 and 499 mm/min. The Z-axis travel is 300 mm and for the X-axis it is 50 mm, resulting in an outer workpiece diameter of 0–90 mm and in an inner diameter of 14–100 mm.

The tool magazine can accommodate three tools (see Fig. 1A) and is rotated by a dc motor. Originally, this was user operated because the controller had no information on the tool that was in the cutting position (active tool). The workpiece is clamped at the chuck and, if needed, at the quill, which is moved by another dc motor, operated, again, by the user (see Fig. 1).

2.2 Control Unit

An overview of the original control circuit is depicted in Fig. 4. This comprises the main power unit, which provides the required voltages of 80 V dc for the main spindle motor, 14 V dc for the tool magazine and quill motors, and 10 V dc for the stepping motor drivers. Figure 4 also depicts: (i) the main signal board for driving the stepping motors: this transforms the G-code commands into suitable pulse trains for each axis; (ii) the circuit corresponding to the drivers of the stepping motors; (iii) the switches for machine start and status indication, spindle start/stop, and emergency stop; (iv) the cooling

Fig. 3 *X*-axis configuration.

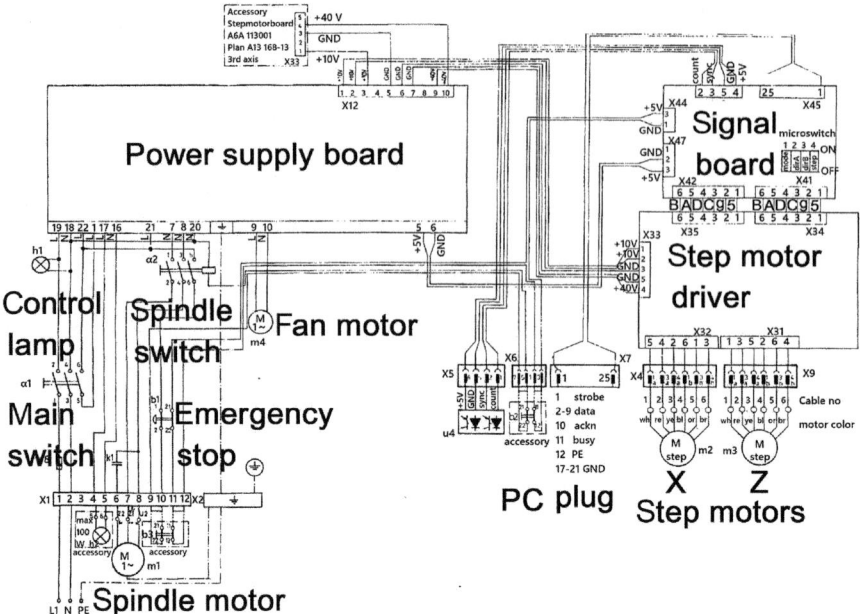

Fig. 4 Machine tool wiring overview.

fan motor for machine electronics; and (v) the available interfaces with external devices. In the original configuration note the existence of a tape reading hardware is appropriately connected to the signal board.

2.3 Operation

The machine possessed originally a complete operation console, see Fig. 5. Controller updating involves keeping only some of the console's elements,

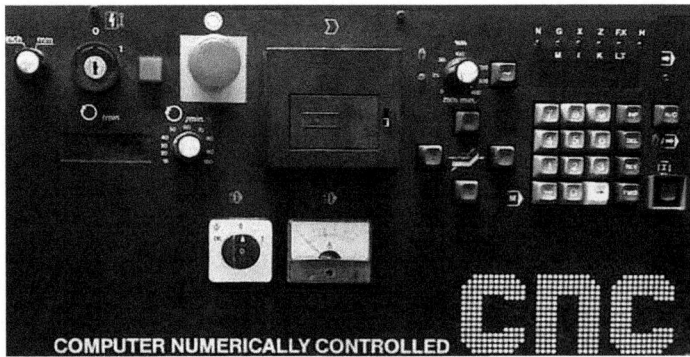

Fig. 5 Original operator console overview.

namely the main switch (0/1), spindle activation switch (0/1/CNC), emergency switch as well as operation indicator lamp, spindle speed indicator, and an amperometer for the spindle motor (see Fig. 1B). All the rest of the switches, keys, knobs, and indicators were replaced by the respective graphical user interface of the new controller.

2.4 Programming

The original programming repertoire was rather outdated by modern standards comprising 36 G-code commands: interpolations (G00, G01, G02/03), hole-making canned cycles (G81–88), unit preparation (G20/21, G90/91, G94/04), thread turning cycles (G33, G78-circular), hole making with chip breaker (G73), radius operation (G24) as well as register command (G92), subroutine call (G25), intermediate transfer (G27), feed motor deactivation (G64), and tape handling (G65). M-codes available comprise standard spindle handling M00/03/05, program handling (M30) and subroutine return (M17), tool change with compensation (M06), task compensation (M98), and circular interpolation mode handling (M99). The resolution originally supported is 0.01 mm.

3 NEW HARDWARE INTERFACES

Both feed and spindle drivers were kept during upgrading. The tape-related electronic circuit was completely removed; so was the signal processing board, since the signals are to be generated by the new soft controller. In addition, new PCBs were created to isolate the feed motor signals as well as those activating the machine and the spindle from the electronic circuits of the PC on which the soft controller runs.

3.1 Main Spindle Speed Control Interface

The soft controller provides the possibility to use a pulse modulated by width (pulse width modulation, PWM) with a frequency varying in proportion to the desired spindle speed. In order to translate this to voltage in the range 0–5 V according to its frequency, a standard circuit comprising npn transistors, capacitors, resistors, and an isolator (PC817) for PC protection was used (see Fig. 6A). Connection was realized through two sockets. The input acquires the signal and ground from the control software through the PC's parallel port (pin 16). The output replaces the potentiometer through which spindle speed control was originally implemented. In the power

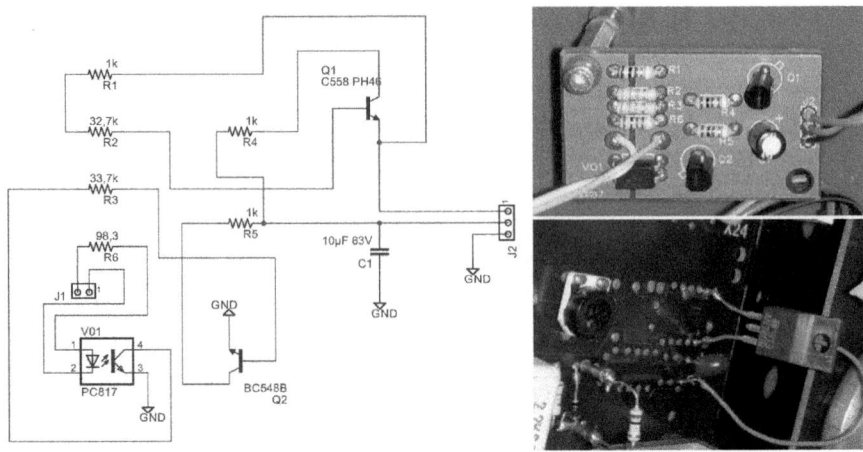

Fig. 6 (A) Circuit transforming PWM to voltage, (B) PWW-DC board, and (C) adaptation of the spindle board.

circuit of the spindle motor, the 27 kΩ resistor was replaced by a voltage regulator (LM 7808) (see Fig. 6C) to avoid constraining the motor current.

3.2 Feed Motor Signal Interface

The feed stepping motor drivers require as input 5 V signals: A, B, C, D, and GND (see Fig. 4) according to quadrature configuration, i.e., for a full step two of the four coils are simultaneously activated in the order AC, BC, BD, AD when motion to one direction is desired, corresponding to phases 1, 2, 3, and 4. When the reverse motion direction is desired, the activation order is simply reversed. Note that phases 1 and 3 are logically opposite, and so are phases 2 and 4. Signals C and D are opposite to A and B, respectively. The designed signal generation circuit is presented in Fig. 7 and constitutes the main part of the feed motor control of the machine tool. A TI LM7875 chip transforms the 14 V of the machine's power supply to constant 5 V fed to the rest of the circuit. Capacitors keep this output voltage independent of any perturbations of the input voltage. The signals obtained from the soft controller through the parallel port of the PC are shown in Fig. 7.

The X-axis is driven by signals in pins 2 and 3. Thus, pin 2 is driven by a resistor at one end (3) of the opto–isolator, whereas the other end (4) of the same side of the isolator is led directly to the ground of the parallel port. When a pulse is present at output 3 of the parallel port, the photodiode of the opto–isolator will emit light and the two opposite pins (5 and 6) of the opto–isolator will conduct, thus a 5 V pulse will be obtained at end 5.

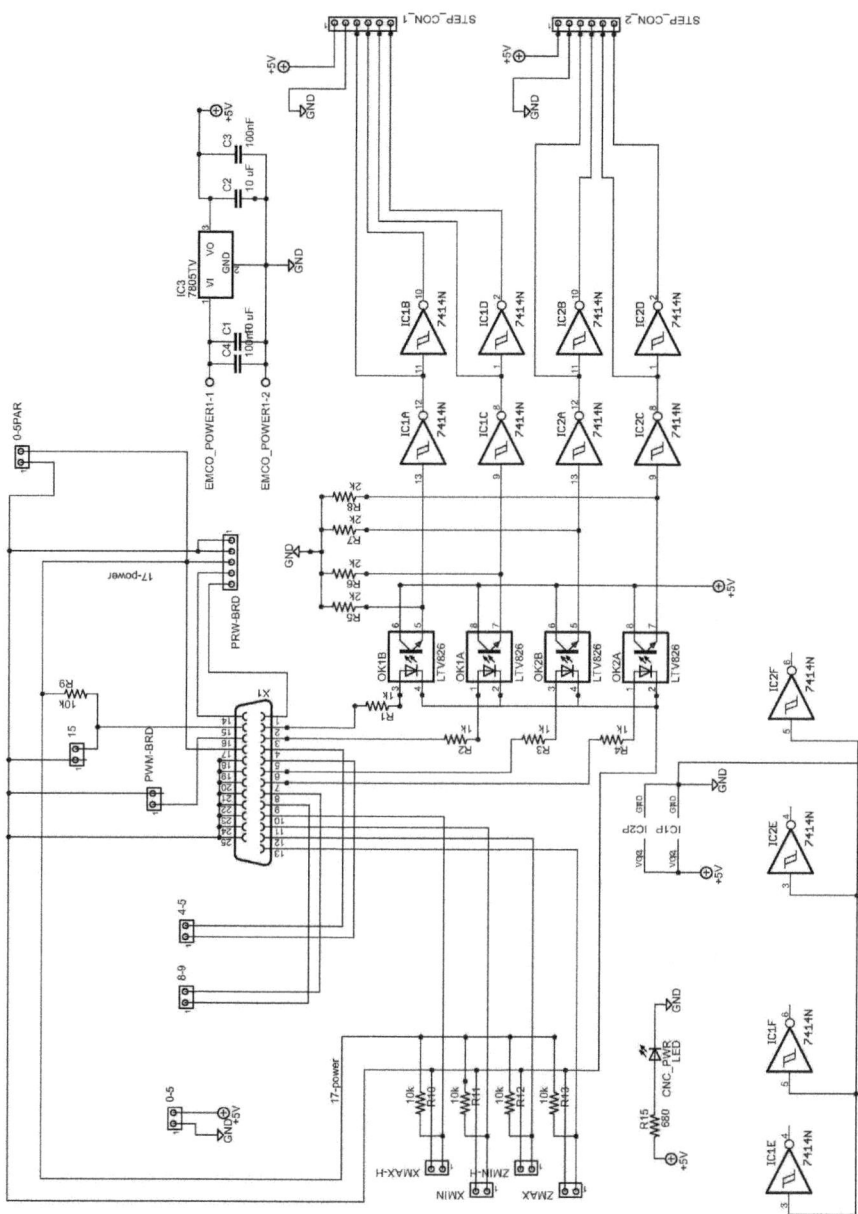

Fig. 7 Input signal circuit to the stepping motors.

In this way, the PC is protected from any short-circuits that may occur in the downstream circuitry. When the photodiode does not emit any light, the signal becomes 0 because a pull-down resistor is used, through which the signal is led to the machine's ground. Subsequently, the signal is driven to an inverter to generate the inverse pulse. Therefore from pin 2 of the parallel port, signals A and C (=not(A)) are generated and from pin 3 signals B and D (=not(B)).

3.3 Limit and Homing Microswitches

The signals driven to the control software are presented in Fig. 7. Input pins 10/11 (12/13) of the parallel port are used in order to pass the information that the X- (Z-)axis has attained its upper/lower limit, according to the respective limit switches. The home switches can be placed anywhere within the respective limits of the axis, but for simplicity it coincides with either of the limit switches. Pull-downresistors of 0 kΩ are used for the connection. An example of placing a microswitch on the machine is shown in Fig. 8. The lower limit switch of the Z-axis is placed on the machine body, i.e., it is fixed in space. The upper limit switch is placed on the tailstock, which is movable.

3.4 Tool Magazine

The rotary tool magazine is shown in Fig. 9A. The motor axis imparts rotation to a ratchet mechanism (see Fig. 9B). The motor rotates continuously so as to block the ratchet thereby keeping the tool in position. At machine

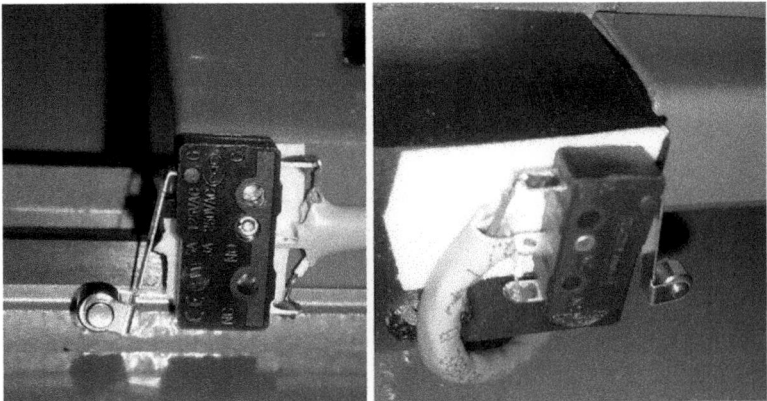

Fig. 8 Limit switches for the Z-axis: (A) upper limit at the tailstock and (B) lower limit on the machine body.

Fig. 9 Tool magazine: (A) overview, (B) motorized ratchet mechanism, and (C) slotted plate constructed.

power-up, the tool magazine rotates until tool no 1 is placed at the cutting position. This information is passed to the controller via an optical sensor collaborating with a slotted disk that is fixed on the tool magazine. The sensor is connected to a simple circuit placed under the slotted disk (see Fig. 9C).

Tool selection is accomplished through three pulse signals of 3.3 V width, one for each tool. Isolation of the respective circuit from the PC's parallel port is achieved by the circuit shown in Fig. 10 using opto–couplers (TI LTV824). In the same circuit, the quill control signal is also shown.

To achieve a tool change, a command is given to the motor to (a) rotate in the opposite direction for 4000/8000 ms, which is the time empirically found to correspond to a rotation angle of 120°/240°, so that the desired

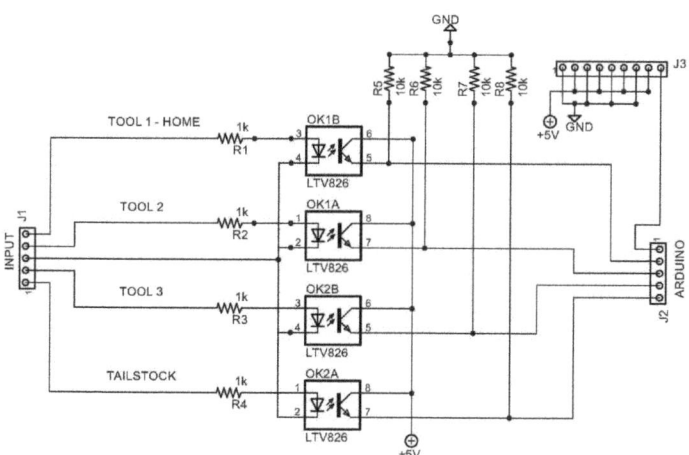

Fig. 10 Circuit for signal connection between the parallel port and the microprocessor.

tool is brought to the cutting position and (b) reverse rotation direction in order to keep the tool in position. This logic is implemented in an Arduino Uno microcontroller fed by 5 V voltage generated by the main board (see Section 3.2). The inputs involved are: (11) the presence of tool 1 at the cutting position (as indicated by the optical sensor), (10) selection of tool 1, (9) selection of tool 2, and (8) selection of tool 3. The pertinent outputs are: (13) the tool magazine rotation in opposite direction to the ratchet and (14) the tool magazine rotation in the same direction to the ratchet. The main program runs continuously, i.e., the inputs are read continuously and a variable is updated every time a different tool is in the cutting position. The output signals of the microcontroller feed a bridge IC (TI L298 Multiwatt-15 type), which implements forward, reverse, fast motor stop, free running, and motor stop states at outputs 3 and 4 (motor pins) depending on the combination of values at digital inputs 3 and 4 (output signals 12 and 13 of the microcontroller). Fig. 11 depicts the circuit designed to control the tool magazine motor.

3.5 Quill

The quill holds the part at the opposite end of the chuck. Its motor is controlled via an output of the parallel port where the respective signal is sent by the soft controller, passing through the opto-coupler (see Fig. 10). The pertinent microcontroller inputs are: (7) hold/release selection, (6) quill movement toward the headstock, and (5) quill movement away from the headstock. The microcontroller reads the parallel port output every 150 ms. A special variable records the current position of the quill. Initially, the quill does not hold any workpiece and the motor moves away from the headstock. When a pulse is generated, the movement is reversed. As in the

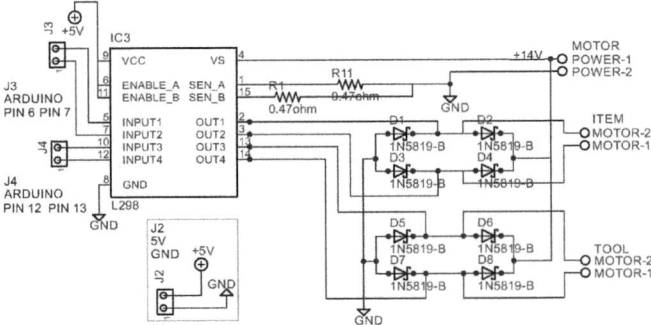

Fig. 11 Control circuit for tool magazine and tailstock quill.

case of the tool magazine, the same bridge IC is used for motor control. Output signals 6 and 7 of the microcontroller, which control the rotation direction of the quill motor, are fed to inputs 1 and 2 of the IC, whereas the motor pins are connected to outputs 1 and 2 (see Fig. 11).

3.6 Power on/off interfaces

Fig. 12 depicts a circuit that was designed to combine powering on of the machine through the console key and the control software's power-on button. In addition, the emergency button is tied to its soft counterpart. This circuit also controls spindle start. This circuit uses three signals of the control software which constitute outputs on the parallel port: (i) emergency button, (2) spindle start, and (17) machine start. These signals are led to opto-isolators and then to a transistor's base. When the base is stimulated a conductive path between the collector and the emitter is formed and, as a result, the core of the corresponding micro-relays K1, K2, and K3 is stimulated at 5 V.

Fig. 13A presents the circuits of the auxiliary contacts of the micro-relays making use of the 230-V line voltage. When the signal on pin 17 of the parallel port is on, then relay K2 is armed and the auxiliary contacts 11–14 και 21–24 close powering the indicator lamp. When the soft emergency stop button is on, i.e., when the signal on pin 1 of the parallel port is high, then relay K1 is armed and contact 11–14 is closed. If the hard emergency button is off and the key-switch is ON, then relay K4 is armed.

Fig. 13B depicts the control circuit for spindle start. When the machine is activated from both the software and the switches and the emergency button is off, then relay K4 is armed, thus contact 11–14 is closed. The spindle is started from the corresponding software button, then relay K3 is armed and contact 11–14 is closed. Contacts MOTOR-1/MOTOR-2 are connected in series with the existing spindle start button; therefore, their start requires the latter to be on.

4 CONTROLLER SOFTWARE INTERFACES

4.1 Open Control Software

LinuxCNC v2.6 (ex "EMC2") was used as an open controller (GNU licensed) running on Linux. It can be configured for various classes of machine tools possessing up to 9 axes. The platform provides: (a) multiple possibilities configurable graphical environment; (b) a G-code

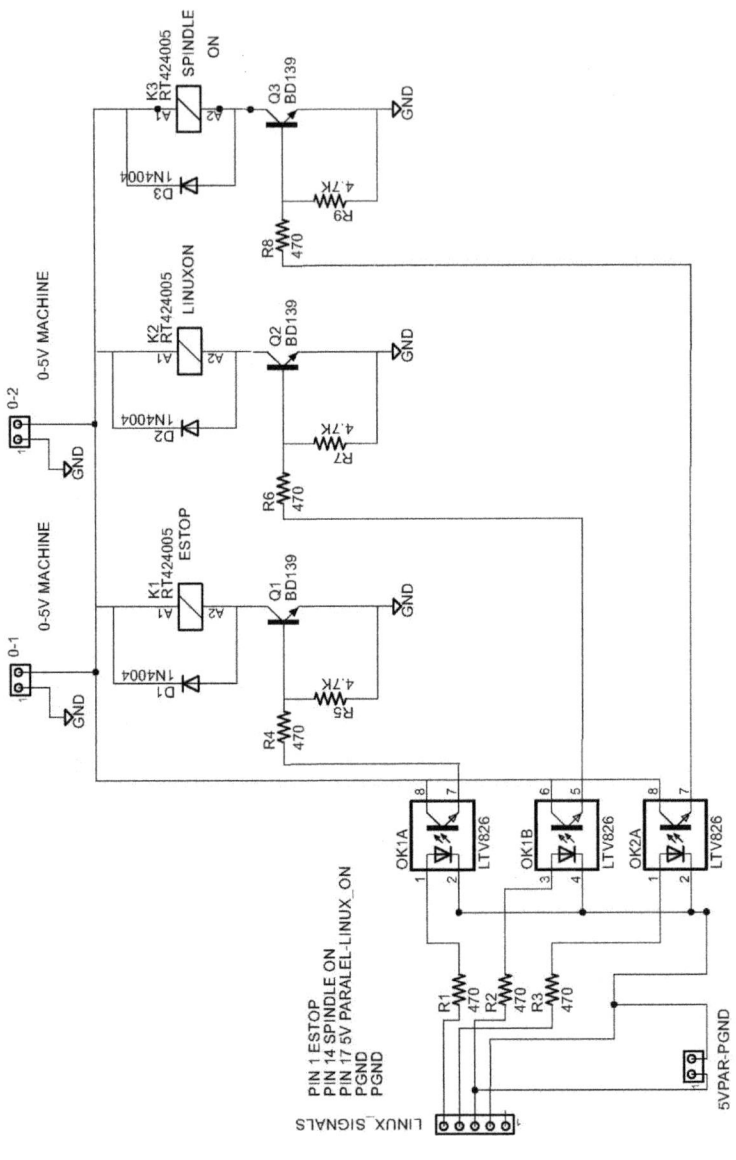

Fig. 12 Machine tool activation circuit.

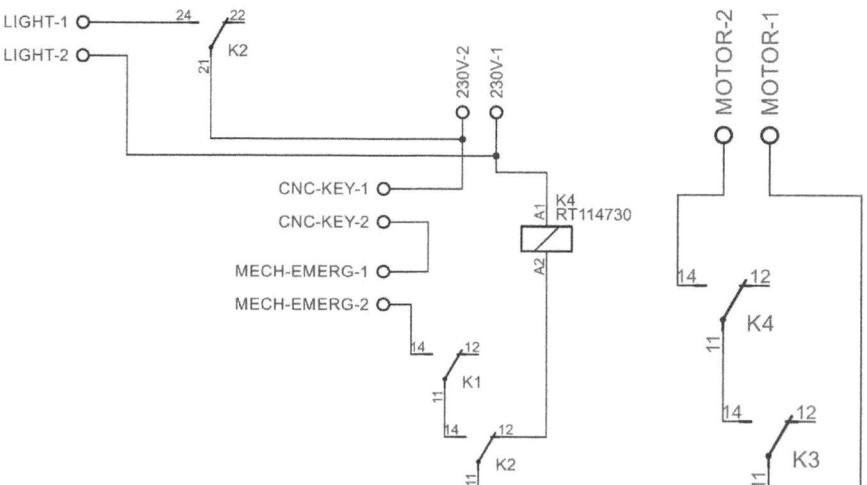

Fig. 13 Control circuits for (A) machine and emergency power on (B) spindle start.

interpreter; (c) a real–time monitoring system for machine tool axes; (d) low–level control of servomotors (analogue or PWM), stepping motors, relays, and sensors; and (e) programmable PLC software with ladder diagrams. The motion control provides for adaptive feed rate as well as continuous speed control. All common operation modes (manual, auto, MDI) are supported. In manual operation, commands are executed by pressing the pertinent buttons on the graphical environment or on the keyboard.

Fig. 14 depicts the GUI as configured for the current application, the pertinent files being (a) *emco_compact_5.ini*, which contains the automatically recorded settings of the axes and spindle (see Sections 4.3 and 4.4); (b) *emco_compact_5.hal*, which records the connection commands with the machine's hardware as generated from the settings (see Sections 4.2 and 3.2); (c) *custom_postgui.hal*, where assignment to custom GUI buttons is made (see Section 4.5); and (d) *custompanel.xml*, where extra GUI elements are recorded and the link to the particular GUI with the respective configuration is put.

4.2 Communications

Table 1 presents the pins of the parallel port of the PC which interfaces the pertinent signals to the machine hardware. Note that pins 18–25 are "grounds."

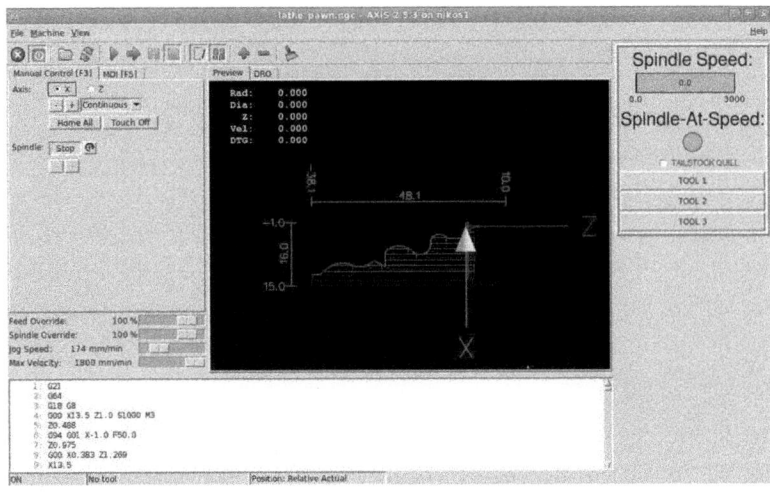

Fig. 14 Configured GUI.

Table 1 Signals assigned to the parallel port pins

Output pin	Type function	Input pin	Type function
1(O)	ESTOP Out	10(I)	Maximum Limit + Home X
2(O)	X Step	11(I)	Minimum Limit X
3(O)	X Direction	12(I)	Minimum Limit + Home Z
4(O)	Digital out 0	13(I)	Maximum Limit Z
5(O)	Digital out 1	14(O)	Spindle CW
6(O)	Z Step	15(I)	Spindle Index
7(O)	Z Direction	16(O)	Spindle PWM
8(O)	Digital out 2	17(O)	Amplifier Enable
9(O)	Digital out 3		

4.3 Axes

X- and Z-axis drivers possess the following characteristics according to their manufacturer: (a) *step time* (pulse duration): 10,000 ns, (b) *step space* (minimum time between successive pulses): 10,000 ns, (c) *direction hold* (duration of keeping a pin signal after its change): 200,000 ns, (d) *direction setup* (duration before a direction change up to the last step pulse): 200,000 ns.

Motion transmission characteristics for both axes are as follows: (a) *steps/revolution*: 72; (b) *microsteps*: 1, meaning that microstepping is not supported; and (c) *motor teeth: 16* and *leadscrew teeth: 40,* effectively defining the transmission ratio. Note that the leadscrew pitch is 2.5 mm/rev.

The above are inputs to forms available in LinuxCNC Stepconf program, which generates a different settings file depending on the machine tool type (see Fig. 15A). In addition, the base address of the parallel port is input as well as the value of *Base Period Maximum Jitter,* which is automatically computed after executing the latency test. The latter refers to the time needed by the PC to stop all other tasks and respond to the external request, i.e., pulse generation and management. For a realistic result, the PC should be overworked for a few minutes.

Furthermore, the right values for velocity and acceleration are sought and the axis test is executed (see Fig. 15B). Initially a low acceleration value is started with, e.g., $50\,\text{mm/s}^2$ and a velocity that is believed to be easily attained within the test area. Acceleration/deceleration should be neither too high, because this may result in step loss of the motors, nor too low, because the axis may overshoot the stopping position. Home location, as well as travel, is defined for each axis, too. Based on the above data some further parameters are calculated by the system such as: the time and distance to reach maximum velocity, the pulse rate for maximum velocity, and the scale of the axis.

The settings for the two limit switches per axis, one of which is also used as homing switch, are shown in Fig. 15C. Home switch location is defined with respect to the machine coordinate system (see Fig. 15A). The home switch is approached by the axis with a velocity equal to the value of the parameter "Home search vel," its sign indicating the retracting direction. If the approach velocity is too large, a large overshoot of the homing position may happen, resulting in destruction of the switch. By contrast, if the approach velocity is too small, the homing process will become lengthy.

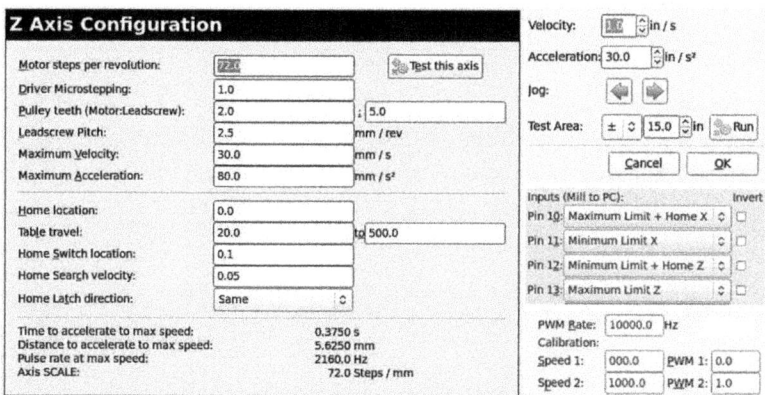

Fig. 15 Settings: (A) axis configuration, (B) velocity-acceleration, (C) switch connections, and (D) spindle PWM.

Parameter "Home latch vel" sets the velocity and direction for precise iden- tification of the homing switch position. Parameter "Home ignore limits" commands the system to ignore limit switches in the homing process. "Home Latch Direction" is set to "same" so that the home position is iden- tified when the switch is closed for the second time, functioning as the limit switch. Homing is done first for the X-axis and then for Z as set by the "Home sequence" parameter.

4.4 Main Spindle

The PWM rate of the machine tool is 10 kHz. The corresponding control setting file use a linear interpolation to calculate the PWM value to be issued for any rotary speed of the spindle in rpm (see Fig. 15D). In MDI mode, two values corresponding to the PWM values 0.1 and 0.8 are commanded by M03, e.g., 100 and 800 rpm, and the recorded values by the machine's encoder as displayed on the digital indicator (see Fig. 1) are inputs to the Stepconf environment. In addition, the spindle speed is depicted online in the GUI form (see Fig. 14).

4.5 GUI

The hardware abstraction layer of LinuxCNC enables the generation of additional GUI windows, buttons, and forms. In this case, this was necessary for tool selection and quill movement. A list with the available tools and their characteristic dimensions (X and Z offsets and tool tip radius) is created (see Fig. 16). The tool magazine may also be operated manually from the GUI (see Fig. 14).

Quill movement (hold/release) may be performed manually (see Fig. 14), but also through an assigned M-code in the range M100 to M199, by creating a text file with the same name, which activates pin 7 of the par- allel port, see Table 1, by issuing the HAL command: "*halcmd setp parport.0. pin-07-out True*".

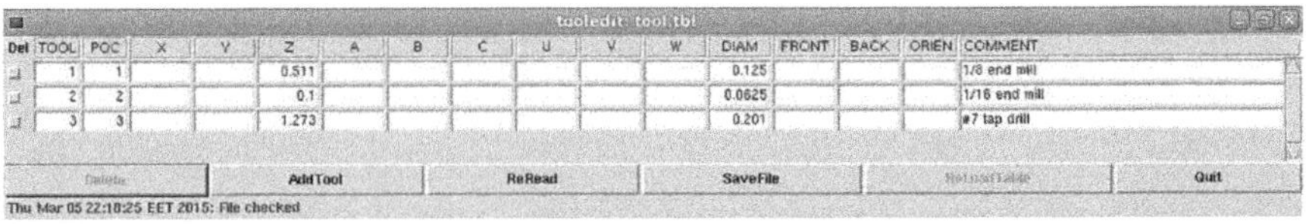

Fig. 16 Tool registration interface.

5 RESULTS, DISCUSSION, AND CONCLUSIONS

Five PCBs were designed using the Eagle software (see Figs. 7 and 10–13) and manufactured on a photosensitive board. Excerpts of the machine configuration file in LinuxCNC are shown in Appendix.1.

The machine tool in its upgraded version was put to test with various workpieces with excellent results for a machine of this size and nature, i.e., dimensional deviations of the order of 10–30 µm. An example is shown in Fig. 17, the corresponding part program is shown in Appendix.2.

An important advantage of open CNC controllers, such as the one that was used in this work, is the ability to match their configuration to the machine configuration and the particular hardware elements. In addition, GUI customization and inbuilt graphical simulation of the movement of axes and of the tools make for operator friendliness. These facilities, together with support of the tested motion control algorithms, the full range of standard commands, and the ability to define custom G and M commands, render open controllers a very appealing alternative when upgrading machine tools. In the machine tool case examined, automatic tool change, spindle speed control, and axis homing became available, whereas in the original specification they were not. In addition, the available G and M commands were extended and custom M commands were added.

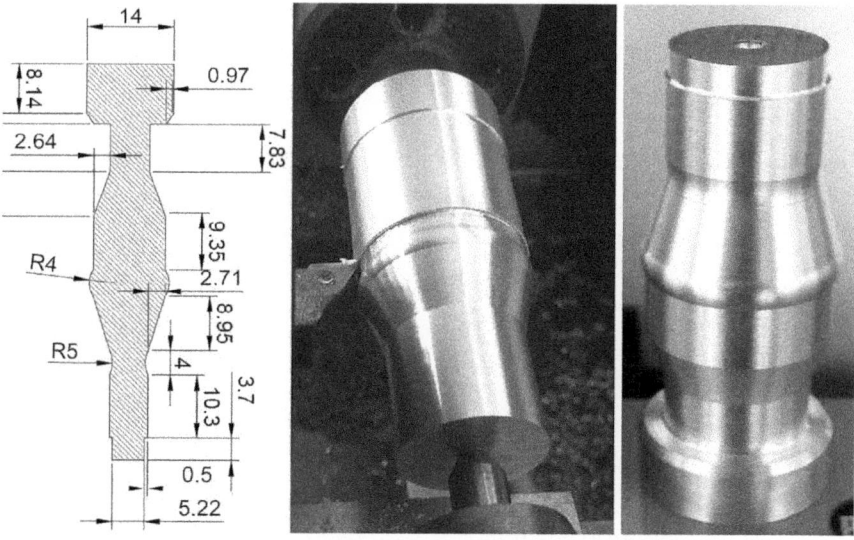

Fig. 17 (A) Sample part drawing, (B) part cutting, and (C) the final form.

In the process of installing an open CNC controller, upgrading of machine tool hardware as such is generally not aimed at. However, the interfacing circuits are needed for the concerning signals to be communicated between the machine and the PC running the software and should be appropriately designed and manufactured. Moreover, there may be other circuits necessary to bypass parts of the original machine control circuitry. In general, modularity of such circuitry is important in this respect.

Due to the openness of the controller, further improvements can be suggested in the particular implementation presented and are relatively easy to materialize. For instance, sensors can be added to each tool slot and a corresponding M code may be defined to wait for tool presence verification. Furthermore, the existing adaptive control option of the controller may be activated in order to modify feed when the spindle current exceeds some limit. Restriction in the number of pins of the parallel port may be overcome by adding a second parallel port in order to assign signals referring to (a) the external devices that are to be controlled through custom M-codes, such as cutting fluid pumps, chip collection conveyor, etc.; (b) monitoring of the subsystems, such as the spindle current, tool condition, etc.; and (c) communication with hierarchical controllers, e.g., of a robot serving the machine, or a manufacturing cell in which the machine is to function, involving other machines, robots, conveyors, etc. Both the work performed and its suggested extension provide an in-depth understanding of the principles and implementation of CNC technology in machine tools, which is nowadays universally applicable in the manufacturing practice and education.

APPENDIX.1 MACHINE CONFIGURATION FILE EXAMPLES (EXCERPTS)

emco_compact_5.ini

......

```
[DISPLAY]
DISPLAY = axis
EDITOR = gedit
....
MAX_FEED_OVERRIDE = 1.2
INCREMENTS = 5 mm 1 mm .5 mm .1 mm .05 mm .01 mm .005 mm
PYVCP = custompanel.xml
....
```

```
[EMCMOT]
EMCMOT = motmod
COMM_TIMEOUT = 1.0
COMM_WAIT = 0.010
BASE_PERIOD = 100000
SERVO_PERIOD = 1000000
[HAL]
HALFILE = emco_compact_5.hal
HALFILE = custom.hal
POSTGUI_HALFILE = custom_postgui.hal
....
[EMCIO]
EMCIO = io
CYCLE_TIME = 0.100
TOOL_TABLE = tool.tbl
[AXIS_0]
TYPE = LINEAR
HOME = 0.0
MAX_VELOCITY = 30.0
MAX_ACCELERATION = 80.0
STEPGEN_MAXACCEL = 100.0
SCALE = 72.0
FERROR = 1
MIN_FERROR = .25
MIN_LIMIT = -10.0
MAX_LIMIT = 40.0
HOME_OFFSET = 0.500000
HOME_SEARCH_VEL = 2.000000
HOME_LATCH_VEL = - 2.000000
HOME_IGNORE_LIMITS = YES
HOME_SEQUENCE = 0
```

emco_compact_5.hal

```
.....
loadrt probe_parport
loadrt hal_parport cfg="0x378 out "
setp parport.0.reset-time 5000
#change of step_type default to define quadrature
loadrt stepgen step_type = 2,2.
#mapping of I/O parallel port pins
```

```
net estop-out => parport.0.pin-01-out
net xstep => parport.0.pin-02-out
net xdir => parport.0.pin-03-out
.....
net toolone => parport.0.pin-04-out
.....
net quill => parport.0.pin-09-out
...
net min-home-z <= parport.0.pin-12-in
net max-z <= parport.0.pin-13-in
net din-00 <= parport.0.pin-15-in
#skipping driver commands with inputs step, dir
#setp stepgen.0.stepspace 0
#setp stepgen.0.dirhold 35000
#setp stepgen.0.dirsetup 35000
setp stepgen.0.maxaccel [AXIS_0]STEPGEN_MAXACCEL
net xpos-cmd axis.0.motor-pos-cmd => stepgen.0.position-cmd
net xpos-fb stepgen.0.position-fb => axis.0.motor-pos-fb
#change of default xstep, xdir to phase-B, phase-A to match step_type
change
net xstep <= stepgen.0.phase-B
net xdir <= stepgen.0.phase-A
net xenable axis.0.amp-enable-out => stepgen.0.enable
net max-home-x => axis.0.home-sw-in
net min-x => axis.0.neg-lim-sw-in
net max-home-x => axis.0.pos-lim-sw-in
```

custompanel.xml
```
<?xml version='1.0' encoding='UTF-8'?>
<pyvcp>
    <vbox>
    <relief>RIDGE</relief>
    <bd>6</bd>
        <label>
            <text>"SpindleSpeed:"</text>
            <font>("Helvetica",20)</font>
        </label>
            ...........................................
            <button>
                <halpin>"quillin"</halpin>
```

```
          <text>"TAILSTOCK QUILL"</text>
     </button>
     <button>
          <halpin>"tool1"</halpin>
          <text>"TOOL 1"</text>
     </button>
     ...........................................
     </vbox>
</pyvcp>
```

custom_postgui.hal

```
.....
#**** Setup of spindle speed display using pyvcp -START ****
#**** Use COMMANDED spindle velocity from LinuxCNC because
no spindle encoder was specified
#**** COMMANDED velocity is signed so we use absolute compo-
nent (abs.0) to remove sign
net spindle-cmd => abs.0.in
net absolute-spindle-vel <= abs.0.out => pyvcp.spindle-speed
#**** force spindle at speed indicator true because we have no feedback
****

#Button connections from custompanel.xml
net spindle-at-speed => pyvcp.spindle-at-speed-led
sets spindle-at-speed true
net toolone => pyvcp.tool1
net tooltwo => pyvcp.tool2
net toolthree => pyvcp.tool3
net quill => pyvcp.quillin
```

APPENDIX.2 TEST PART PROGRAM%

```
N5 G40 G90 G54 G21 G18
N10 S1000 M03 T01
N15 G28
N16 G00 X0 Z0
N25 G01 X1 F70
N30 G00 Z68
N40 G01 X-0.5 F70
N50 Z-0.5
```

```
N60 X1
N61 G00 Z68
N62 G01 X-1 F70
N63 Z8.86
N64 X-0.5 Z8.1368
N65 X2
N66 Z18.043
N67 X-1
N68 X-1.3 Z17.155
N69 Z9.33
N70 X-0.5 Z8.1368
N71 X2
N80 Z19.5166
N85 X-1
N90 X-1.82 Z17.24
N95 Z9.85
N100 X-1.676
N110 X-1.32 Z9.33
N115 G04 P10
N120 X2
N130 Z20.999
N140 X-1
N150 X-2.32 Z17.33
N160 Z9.85
N170 X2
N180 Z22.464
N190 X-1
N200 X-2.82 Z17.42
N210 Z9.85
N220 X2
N230 Z23.94
N240 X-1
N250 X-3.32 Z17.505
N260 Z9.85
N270 X2
N275 G04 P10
N280 Z25.41
N290 X-1
N300 X-3.82 Z17.59
```

```
N310 Z9.85
N320 X2
N330 Z26.89
N340 X-1
N350 X-4.32 Z17.68
N360 Z9.85
N370 X2
N380 G00 Z68
N385 G04 P10
N390 G01 X-1.8 F70
N400 Z47
N410 G01 X-1.68 Z38.05
N420 G00 Z68
N430 G01 X-2.6 F70
N440 Z47
N450 G01 X-1.68 Z38.05
N460 G00 Z68
N470 G01 X-3.5 F70
N480 Z47
N490 G01 X-1.68 Z38.05
N500 G00 Z68
N510 G01 X-4.39 F70
N520 Z47
N530 G01 X-1.68 Z38.05
N540 G03 X-1.68 Z33.72 R4
N550 G01 Z25 F70
N560 X2
N570 G00 Z68
N580 G01 X-4.89 F70
N590 Z61.3
N600 X0
N610 G00 Z0
N620 G28
N625 G04 P10
N630 M06 T02 G43
N640 M03 S1200
N650 G00 Z51
N660 G01 X-4.39 F70
N670 G02 X-4.38 Z47 R5
N680 G01 X2 F70
```

N690 G28
N700M05
N710 M30
%

REFERENCES

Du, Y., Cao, H., Liu, F., et al., 2012. An integrated method for evaluating the remanufacturability of used machine tool. J. Clean. Prod. 20, 82–91.

Fei, J., Deng, R., Zhang, Z., et al., 2011. Research on embedded CNC device based on ARM and FPGA. Procedia Eng 16, 818–824.

Gutierrez, M.E., Álvares, A.J., 2013. Development of a Cnc router adherent to standard step-Nc based on the controller advanced machine (Emc2). 22nd International Congress of Mechanical Engineering. Ribeirão Preto, Brazil, pp. 8200–8213.

Hentz, J.B., Nguyen, V.K., Maeder, W., et al., 2013. An enabling digital foundation towards smart machining. Procedia CIRP 12, 240–245.

Ji, H., Li, Y., Wang, J., 2008. A software oriented CNC system based on Linux/RTLinux. Int. J. Adv. Manuf. Technol. 39, 291–301.

Khanna, A., Kumar, A., Bhatnagar, A., et al., 2013. Low-cost production CNC system. 7th International Conference on Intelligent Systems and Control (ISCO). IEEE, pp. 523–528.

LinuxCNChttpLinuxCNChttp://www.linuxcnc.org/ (2016, accessed 23 February 2017).

Ma, X., Han, Z., Wang, Y., et al., 2007. Development of a PC-based open architecture software-CNC system. Chin. J. Aeronaut. 20, 272–281.

Martinov, G.M., Lyubimov, A.B., Bondarenko, A.I., et al., 2015. An approach to building a multiprotocol CNC system. Autom. Remote. Control. 76, 172–178.

Park, S., Kim, S.-H., Cho, H., 2006. Kernel software for efficiently building, re-configuring, and distributing an open CNC controller. Int. J. Adv. Manuf. Technol. 27, 788–796.

Pritschow, G., Altintas, Y., Jovane, F., et al., 2001. Open controller architecture–past, present and future. CIRP Ann Manuf Technol 50, 463–470.

Staroveški, T., Brezak, D., Udiljak, T., et al., 2009. Implementation of a Linux-based CNC open control system. 12th International Scientific Conference on Production Engineering–CIM, Biograd, Croatia, pp. 209–216.

Staroveski T, Brezak D, Udiljak T, et al. Experimental machine tool for process monitoring and control systems research. In: Annals of DAAAM and Proceedings of the International DAAAM Symposium. 2011, pp. 23–24.

Uhlmann, E., Lang, K.-D., Prasol, L., et al., 2017. Sustainable solutions for machine tools. In: Sustainable Manufacturing: Challenges, Solutions and Implementation Perspectives, Part II. Springer International Publishing, pp. 47–69.

Wang, T., Wang, L., Liu, Q., 2011. A three-ply reconfigurable CNC system based on FPGA and field-bus. Int. J. Adv. Manuf. Technol. 57, 671–682.

Wang, P., Liu, Y., Ong, S.K., et al., 2014. Modular design of machine tools to facilitate design for disassembly and remanufacturing. Procedia CIRP 15, 443–448.

Xu, W., Chen, J., 2012. Research on ARM numerical control system. Phys. Procedia 25, 1934–1938.

Xu, X., Li, Y., Sun, J., et al., 2012. Research and development of open CNC system based on PC and motion controller. Procedia Eng 29, 1845–1850.

Zhang, C., Wang, H., Wang, J., 2003. An USB-based software CNC system. J. Mater. Process. Technol. 139, 286–290.

Zhang, X.L., Tang, X.Q., Chen, J.H., et al., 2007. Hierarchical real-time networked CNC system based on the transparent model of industrial Ethernet. Int. J. Adv. Manuf. Technol. 34, 161–167.

CHAPTER 6

To Model the Assembly of Thin Parts in Composite Material

Wilma Polini, Andrea Corrado
Department of Civil and Mechanical Engineering, Università di Cassino e del Lazio Meridionale, Cassino, Italy

1 INTRODUCTION

An assembly consists of two or more components or subassemblies. Owing to variations in manufacturing, it is impossible to completely avoid deviations in a component's dimensions and geometry. Tolerance analysis enables the prediction of the effects of these part deviations on assembly functional requirements.

Historically, tolerance analysis has always been divided in two kinds of analysis: an analysis involving rigid parts and an analysis involving compliant parts in the assembly.

A substantial number of mathematical models for tolerance analysis of rigid assemblies have been proposed, i.e., virtual boundary (Jayaraman and Srinivasan, 1989), variation model (Gupta and Turner, 1993), TTRS model (Desrochers, 2003), matrix model (Desrochers and Rivière, 1997), vector loop (Gao et al., 1998), Jacobian model (Laperrière and Lafond, 1999), torsor model (Bourdet et al., 1996), Jacobian-torsor model (Ghie et al., 2003), T–Map model (Davidson et al., 2002), deviation domain (Giordano et al., 1999), and skin model (Schleich et al., 2014). Moreover, many software packages exist and allow making the tolerance analysis of rigid assemblies. These software packages such as MECAMaster, Sigmund, 3DCS, VisVSA, CeTol, and PolitoCAT are based on some approaches cited previously.

These methods are inadequate when parts exhibit a flexibility (Corrado and Polini, 2017a; Corrado and Polini, 2017b). The elastic flexibility of such parts may cause wide shape variations during assembly process that combines with tolerances on parts and fixtures, so causing uncertainty to predict the actual shape configuration of the final assembly product once released by manufacturing fixtures.

Manufacturing Engineering Education
https://doi.org/10.1016/B978-0-08-101247-5.00006-X
131

For these reasons, many researchers proposed several methodologies mainly based on FEM to analyze the behavior of compliant part assemblies. The first researches proposed a methodology called the Place, Clamp, Fasten and Release (PCFR) cycle; it was inspired by the cycle used to assemble parts or subassemblies of automobile body at each station along the line of handling (Chang and Gossard, 1997). Other researchers extended the Vector-Loop method from rigid to compliant assemblies and developed a general framework, called FASTA (Flexible Assemblies Statistical Tolerance Analysis) that allowed the static prediction of the variability of flexible assemblies from experimental measurements (Mortensen, 2002). The milestone in the field of tolerance analysis of compliant assemblies is the methodology proposed in Liu and Hu (1997), that it was able to simulate compliant part assemblies based on the concept of influence coefficients. The method of influence coefficient (MIC) was then extended to multi-station processes involving fixturing and tooling deviations (Camelio et al., 2003) and to the geometric covariance (Camelio et al., 2004). It was based on linear assumption. However, linear approaches did not provide adequate results when large deformations occur, or part-to-part contact conditions have to be taken into account. In Xie et al. (2007) how contact among parts being assembled highly influences final assembly shape was shown. These methods are more accurate than a linear one, but they are very time consuming especially if combined with Monte Carlo-based simulations. To overcome this lack, a linear contact algorithm was proposed both in Ungemach and Mantwill (2008), by combining a linear contact search and a contact equilibrium criterion. Simulation of geometrical variation in injection–molded plastic components has been discussed in Lorin et al. (2012). In Lorin et al. (2013) and Jareteg et al. (2014), the authors proposed a method for analyzing thermal expansion in combination with geometrical variation and an integration of manufacturing process simulations of composite materials into the field of variation analysis. Also in this case, some software packages exist: TAA (Sellem and Rivière, 1998), 3DCS-FEA, VisVSA-FEA, RD&T, ANATOLEFLEX (Falgarone et al., 2016), and SVA-FEA (Franciosa et al., 2011).

The entire literature has focused on the assembly of metal sheets by welding or riveting and considering the linear and nonlinear phenomena due to thermal effects or residual stresses during assembly. However, other materials that are typically compliant are used, such as composite materials, in which the assemblies are not made by welding but by means of mechanical fixing elements (bolting, riveting) and/or structural adhesives. Moreover, the works of the literature did not consider the critical aspect of the simulation time, apart from the MIC which has a simulation time smaller than that of a FEM approach. The computational efficiency to simulate the compliant assemblies depends on the discretization of the bodies inside FEM software. Mesh size plays a significant role for the accuracy of the analysis; in fact, the smaller the elements more are the steps needed to model actual part behavior. Moreover, the smaller the elements more are the nodes and the computational time involved. However, the problem of computational efficiency remains, even if few elements are chosen.

To solve this problem, a way should be found to reduce the simulation time by considering the same number of elements used to discretize the geometry of the parts to assembly. This way is described in this work that is organized as follows: in Section 2, a description of the most used methods to solve tolerance analysis problems of compliant assemblies is discussed in depth. In Section 3, the new method to simulate compliant assemblies is shown. Section 4 presents a case study, constituted by an assembly of thin parts in composite materials joined by an adhesive, in order to underline the difference in terms of simulation times between the new method and the approach of the literature.

2 TOLERANCE ANALYSIS OF COMPLIANT ASSEMBLY

The most widely used method of the literature to solve tolerance analysis problems is based on the method of influence coefficient (MIC). This method divides the assembly process into four steps, as shown in Fig. 1.

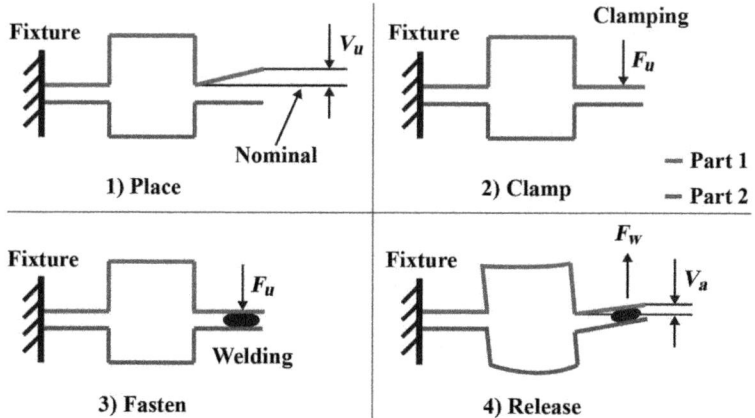

Fig. 1 A schematic representation of a compliant assembly process.

Step 1 shows the part variation V_p when the parts are loaded onto the workholding fixture. If more than one source of variation is considered, the variation of parts is expressed by a vector $\{V_p\}$. In Step 2, clamps are applied to force the parts to their nominal positions. Finite element modeling (FEM) is used to calculate the forces applied by the clamps. The linear relationship between forces and displacements may be written as:

$$\{F_p\} = [K_p] \cdot \{V_p\} \tag{1}$$

where $\{F_p\}$ is the force vector, $[K_p]$ is the stiffness matrix of the parts, and $\{V_p\}$ is the deviation vector, resulting from part variation at each node. The components of the vector $\{F_p\}$ represent forces provided by clamps. Items with subscript p represent quantities of the unjoined parts. In Step 3, parts are joined to form the assembly; thus, the stiffness of the joined parts increases from $[K_p]$ to $[K_a]$. Items with subscript a representing quantities of the joined parts. In Step 4, the clamps are released from the joined assembly. The assembly will spring back from its nominal position after the forces of the clamps are released. FEM is used one more time to calculate the spring back of the assembly:

$$[K_a] \cdot \{V_a\} = \{F_a\} \quad \text{or} \quad \{V_a\} = [K_a]^{-1} \cdot \{F_a\} \tag{2}$$

where $\{V_a\}$ is the spring back of the assembly at each node, and $\{F_a\}$ is the force vector.

The key to the method is to establish a linear relationship between the part deviations $\{V_p\}$ and the assembly spring–back deviations $\{V_a\}$. Under the small displacement hypothesis $\{F_p\} = \{F_a\}$; then:

$$[K_a] \cdot \{V_a\} = [K_p] \cdot \{V_p\} \rightarrow \{V_a\} = [K_a]^{-1} \cdot [K_p] \cdot \{V_p\} = [S] \cdot \{V_p\} \tag{3}$$

where $[S]$ is the sensitivity matrix and it is equal to the product of the matrices $[K_a]^{-1}$ and $[K_p]$. The method of influence coefficients helps to evaluate the sensitivity matrix by the following steps. In the first step, a unit force is applied at the ith node of geometry where there is a variation ($i = 1$ to n). The direction of the unit force is the same as the variation direction. FEM is used to calculate the displacement under the i-th unit force in the n nodes interested by geometry variations (input). The displacements are recorded into a matrix, called $[C]$, that represents the matrix of influence coefficients. At this point, the stiffness matrix, $[K_p]$, is just the inverse of the matrix of influence coefficients. The stiffness matrix will be an $n \times n$ matrix with n equal to the number of part variations that are the input of the tolerance analysis. Therefore, the evaluation of the stiffness matrix $[K_p]$ requires n runs of FEM simulation.

In the second step, FEM is used one more time to calculate the spring-back deviations at any nodes in any directions for the assembly. The ith column of matrix $[K_p]$ ($i = 1$ to n) is treated as the forces and it is applied on the nodes interested by geometry variations and the obtained displacements are recorded in the matrix $[S]$. To pass from the stiffness matrix $[K_p]$ to the sensitivity matrix $[S]$ n further runs of FEM simulation are needed. So, the whole matrix $[S]$ is evaluated by $2n$ FEM runs through the method of influence coefficients.

The assembly can be described using the mean deviations and the variance of the assembly characteristic points; the mean deviation vector $\{\mu_a\}$ of compliant assembly can be calculated as:

$$\{\mu_a\} = [S] \cdot \{\mu_p\} \tag{4}$$

where $\{\mu_p\}$ is the mean deviation vector of the part variations $\{V_p\}$. In addition, the assembly covariance matrix $[\sum_a]$ can be calculated as

$$\left[\sum\nolimits_a\right] = [S] \cdot \left[\sum\nolimits_p\right] \cdot [S^T] \tag{5}$$

where $[\sum_p]$ represents the covariance matrix for the parts. If the part variations are statistically independent, the assembly covariance matrix $[\sum_a]$ is expressed in terms of the variance vector $\{\sigma_a^2\}$:

$$\{\sigma_a^2\} = [S^2] \cdot \{\sigma_p^2\} \tag{6}$$

where $\{\sigma_p^2\}$ is the variance vector of the part variations, $\{V_p\}$. The standard deviation of the assembly $\{\sigma_a\}$ can be calculated by taking the square root of Eq. (6).

3 THE MODIFIED METHOD OF INFLUENCE COEFFICIENTS (MMIC)

The aim of the proposed approach is to reduce the simulation time involved by MIC. To do this, it is necessary to take action on the number of FEM simulations. The most widely used method to solve tolerance analysis problems, explained previously, can be modified to achieve the aim. In fact, the step that requires more time is related to the evaluation of the stiffness matrix $[K_p]$ because, due to the inability to assess this matrix within the FEM explicitly because it was internally formulated, the authors were forced to use the method of influence coefficients.

Nowadays, many FEM solvers provide the stiffness matrix explicitly to the user. However, it can be still very difficult for the user to manipulate the matrix, because as the number of elements and the degree of freedom of nodes increase so does the size of the stiffness matrix. Therefore, it is necessary to reduce the matrix size without loss of accuracy.

This aspect has already been solved by using the super-element method in finite element analysis (Crisfield, 1986). This method reduces the degrees of freedom in the stiffness matrix to only those degrees of freedom that exist at defined boundary nodes in order to reduce the stiffness matrix without loss of accuracy. The boundary nodes include fastening points, locations of applied forces, and nodes with applied constraints. All other nodes are referred to as interior nodes and are important only because of their influence on the boundary nodes. The interior details are removed, but the essential elastic behavior is retained. The matrix can be sorted and partitioned as shown in Eq. (7).

$$[K_p] = \begin{bmatrix} \overbrace{\text{DOF on boundary}} & \overbrace{\text{Coupled DOF}} \\ K_{aa} & K_{ba} \\ K_{ab} & K_{bb} \\ \underbrace{\text{Coupled DOF}} & \underbrace{\text{DOF Interior}} \end{bmatrix} \tag{7}$$

where DOF on boundary stands for the degree of freedom of fastening or constrained nodes (nodes of boundary conditions), coupled DOF contains terms that couple the boundary and the interior nodes and DOF interior is constituted by the degree of freedom of interior nodes. Partitioning the force and displacement vectors in the same manner, the linear stiffness equation can be written as

$$\begin{Bmatrix} F_{aa} \\ F_{bb} \end{Bmatrix} = \begin{bmatrix} K_{aa} & K_{ba} \\ K_{ab} & K_{bb} \end{bmatrix} \cdot \begin{Bmatrix} \delta_{aa} \\ \delta_{bb} \end{Bmatrix} \tag{8}$$

where F_{aa} and δ_{aa} represent the forces and the displacements on the boundary, and F_{bb} and δ_{bb} represent the forces and the displacements on the interior. If there are no forces on the nodes, this expression can be rewritten as

$$F_{aa} = K_{aa} \cdot \delta_{aa} + K_{ba} \cdot \delta_{bb} \tag{9}$$

$$0 = K_{ab} \cdot \delta_{aa} + K_{bb} \cdot \delta_{bb} \tag{10}$$

Eq. (10) can be solved for δ_{bb} and the result can be substituted into Eq. (9) as follows:

$$F_{aa} = \left(K_{aa} - K_{ba} \cdot K_{bb}^{-1} \cdot K_{ab} \right) \cdot \delta_{aa} \qquad (11)$$

The super-element stiffness matrix is the term in the parenthesis of Eq. (11) and relates the boundary forces to the boundary displacements. The stiffness matrix evaluated with the super-element method provides some advantages. It is smaller than the global stiffness matrix; its size depends only by the number of degrees of freedom associated with the boundary; then, the smaller size reduces the computation time required for manipulating large matrices.

The architecture of the proposed approach is based on Matlab to perform computations with matrices and vectors and MSC Marc solver to evaluate the stiffness matrix. MSC Marc software is able to provide the stiffness matrix explicitly to the user by using the super-element method but also to evaluate this matrix, according to the method of influence coefficients, with the help of a script and a subroutine that automatically execute the required simulation runs.

The differences between the proposed approach and the method explained above are shown in Fig. 2. With the proposed approach, the sensitivity matrix $[S]$ is evaluated with $(1 + n)$ FEM runs, rather than $2n$ FEM runs, due to the stiffness matrix $[K_p]$ that requires only one FEM run by the super-element method.

To verify that the proposed approach really allows reducing simulation times, a case study was taken into account and solved with both the classic and proposed approaches.

4 TOLERANCE ANALYSIS OF THIN PART ASSEMBLY IN COMPOSITE MATERIAL

4.1 Case Study

The case study is a T-shaped assembly made up of three parts, as shown in Fig. 3. This is a type of structure commonly used in the design of components in composite material. All parts are joined by an adhesive and only L-shaped parts have an imposed geometrical error called spring-in. This geometric error is common to parts made of composite material and is widely studied (Ersoy et al., 2010; Kappel et al., 2013).

The tolerance analysis of the case study has been carried out according to the workflow shown in Fig. 4. Starting from the nominal assembly geometry, the FEM model is created accordingly and the shell elements (QUAD element 75) are used. Once input data are correctly assigned, output data, in terms of mean variations and standard deviations, are given by solving the fastening process and the assembly sequence.

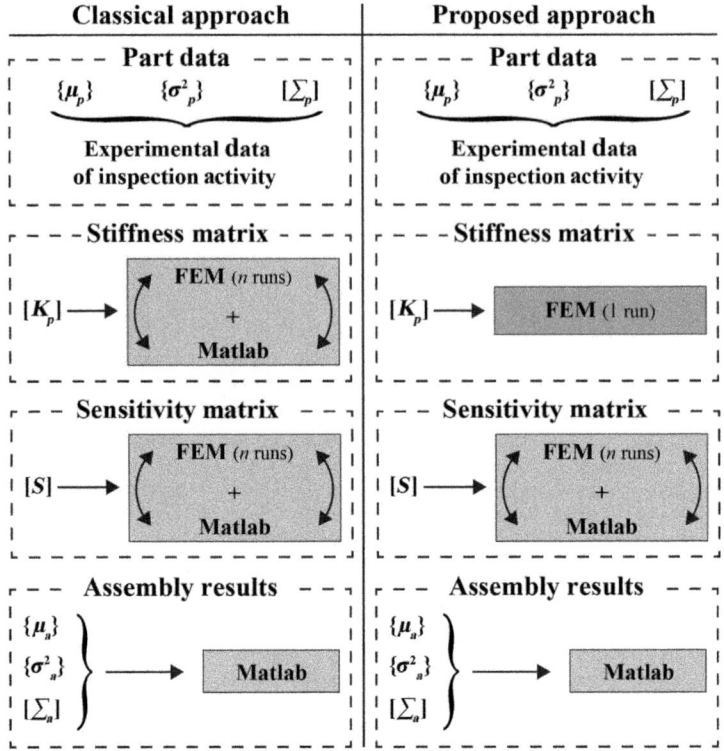

Fig. 2 Schematic differences between two approaches.

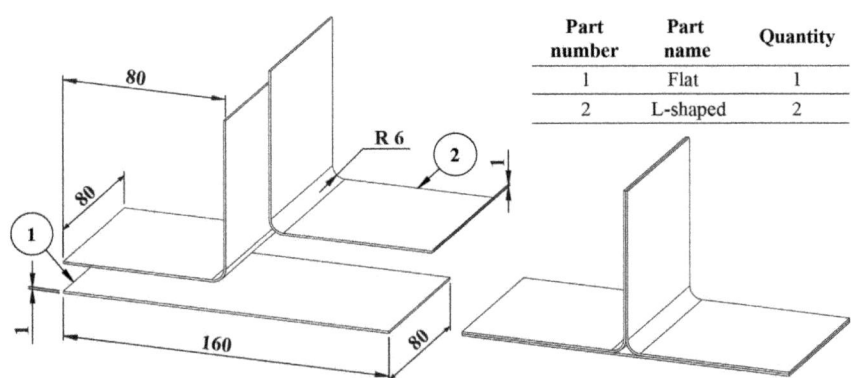

Fig. 3 T-shaped assembly (dimensions in mm).

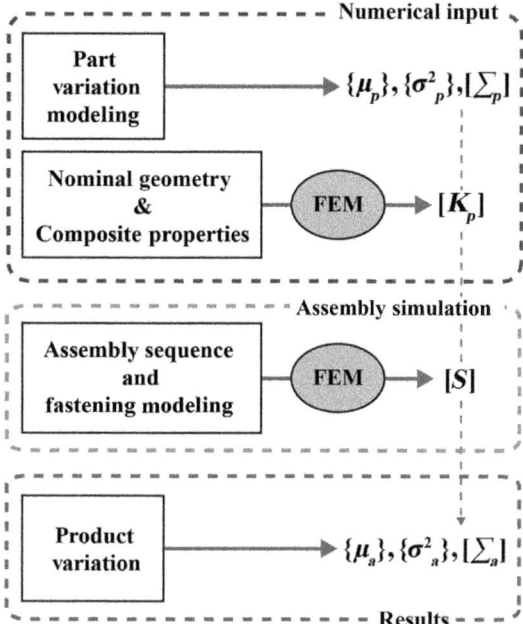

Fig. 4 Numerical workflow.

4.2 Step 1: Numerical Input

The case study was modeled by a mapped meshing with a total number of nodes and elements equal to 1751 and 1600, respectively (see Fig. 5). All parts are made up of unidirectional prepreg material with [0, 90, 0 degrees]$_\text{S}$

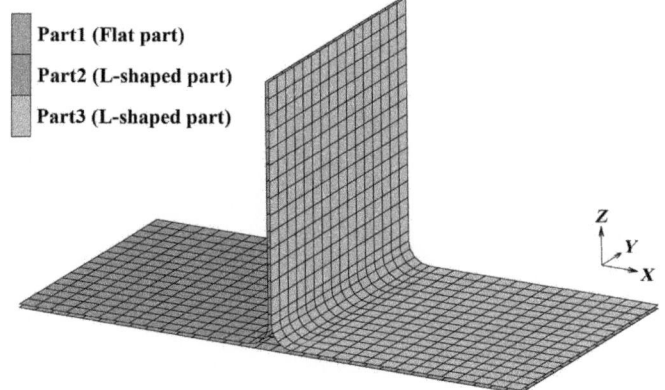

Fig. 5 The mapped meshing of the case study.

Table 1 Material properties of Cycom970/T300

Property	Value
E_1 (MPa)	120,000
E_2 (MPa)	8000
E_3 (MPa)	8000
ν_{12}	0.30
ν_{23}	0.40
ν_{31}	0.02
G_{12} (MPa)	5000
G_{23} (MPa)	2760
G_{31} (MPa)	5000

as lay-up sequence. The nominal material properties of a single ply are shown in Table 1.

The mean value of spring-in and its standard deviation are equal to 1.220 and 0.080 degrees, respectively, for the L-shaped parts. This angular error has to be translated into a linear variation using trigonometric relations. As a matter of fact, considering the input variation applied to the share of 74.5 mm and knowing the spring-in, the triangle cathetus can be determined and then the input variation in terms of mean and variance can be defined, as shown in Fig. 6A. In particular, the variation with a mean and standard deviation of 1.586 and 0.104 mm, respectively, will be applied only on three nodes for each part (nodes 860, 950, and 1040 for part 2; and nodes 1455, 1545, and 1635 for part 3), as shown by the red circles in Fig. 6B. Therefore, the mean variation and the variance vectors of the part variation sources are:

$$\{\mu_p\} = \left\{ \overbrace{-1.586 \ -1.586 \ -1.586}^{\text{Part2}} \ \overbrace{1.586 \ 1.586 \ 1.586}^{\text{Part3}} \right\}^{\text{T}} \tag{12}$$

$$\{\sigma_p^2\} = \left\{ \overbrace{0.011 \ 0.011 \ 0.011}^{\text{Part2}} \ \overbrace{0.011 \ 0.011 \ 0.011}^{\text{Part3}} \right\}^{\text{T}} \tag{13}$$

The first three values of the vector $\{\mu_p\}$ have the minus sign to respect the direction of the variation applied on the three points of part 2 as regards to the reference system taken into account. This aspect can also be explained by the fact that the L-shaped parts, which have the same geometric error, are placed on the flat part in the opposite direction, as shown in Fig. 3. To evaluate the stiffness matrix $[K_p]$, it was necessary to constrain the geometry

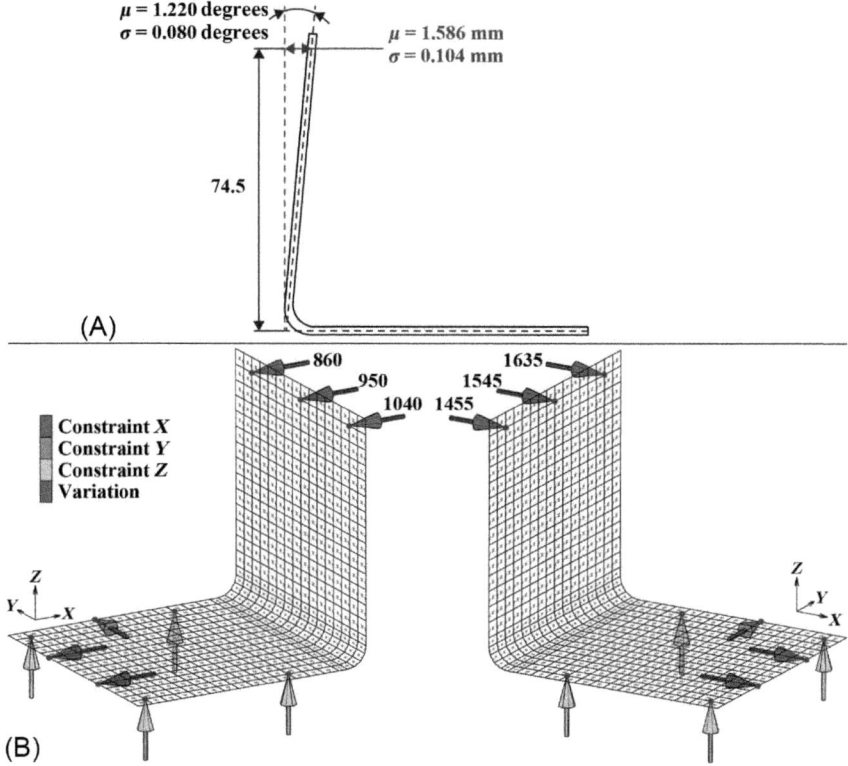

$\mu = 1.220$ degrees
$\sigma = 0.080$ degrees

$\mu = 1.586$ mm
$\sigma = 0.104$ mm

74.5

(A)

860
950
1040   1455
1545
1635

Constraint X
Constraint Y
Constraint Z
Variation

(B)

Fig. 6 (A) Linear input variation and (B) place where to apply variations and constraints.

affected by the geometric variation, as shown in Fig. 6B. Therefore, the stiffness matrix was evaluated with both the classic and proposed approaches, as explained previously.

4.3 Step 2: Assembly Simulation

In order to evaluate the sensitivity matrix [S], it was necessary to constrain the entire assembly. These constraints represent the fixturing system of the parts during the assembly and release steps, as shown in Fig. 7. But, these constraints are not enough to evaluate the sensitivity matrix. As a matter of fact, it is also necessary to indicate how to perform the fastening of the parts. The assembly cycle foresees that all parts are initially separated and then they are assembled simultaneously. The fastening operation consists in applying a point-to-point connection, by means of beam elements, in all areas of assembly between all nodes of parts (see Fig. 8). The point-to-point connection was used by the CWELD connection method in order to

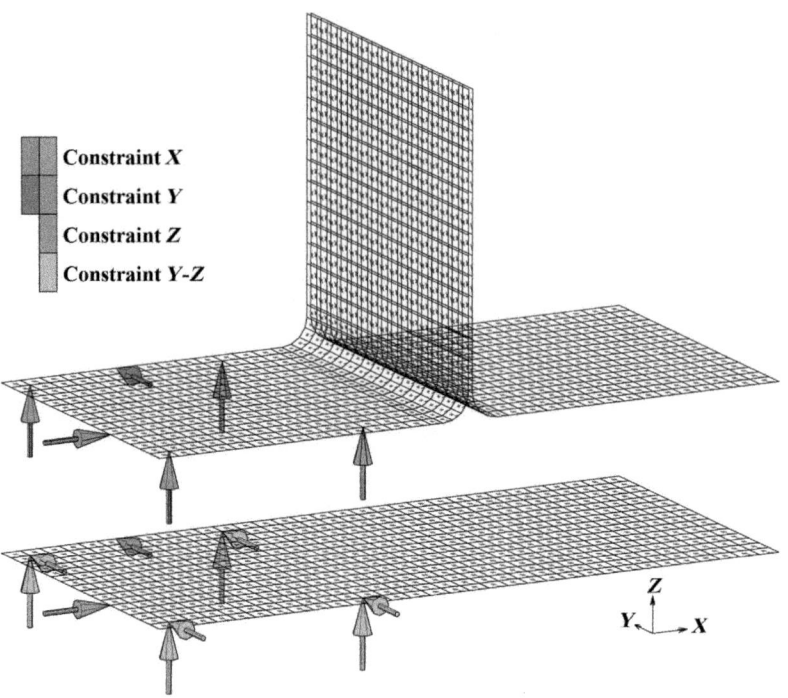

Fig. 7 Position of constraints for the T-shaped assembly during the assembly and release steps.

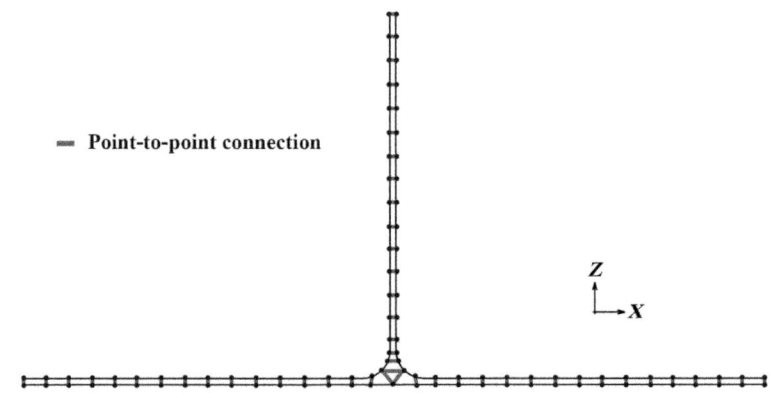

Fig. 8 Fastening method (section view).

simulate the action of glue (Marc, 2013). The beam elements were used with a Young Modulus and a Poisson coefficient of, respectively, 1500 MPa and 0.4, and a solid square cross-section with a side of 5 mm. The mechanical properties refer to ADEKIT A140 made by Axson.

4.4 Step 3: Results and Analysis

Once the sensitivity matrix is evaluated, the assembly variations are evaluated starting from Eqs. (4) and (6) in order to have the mean variation $\{\mu_a\}$ and the standard deviation $\{\sigma_a\}$.

The mean variation and standard deviation of the assembly are shown in Fig. 9. The mean variation and standard deviation of the assembly become less than those of the parts. The maximum values of the mean variation and

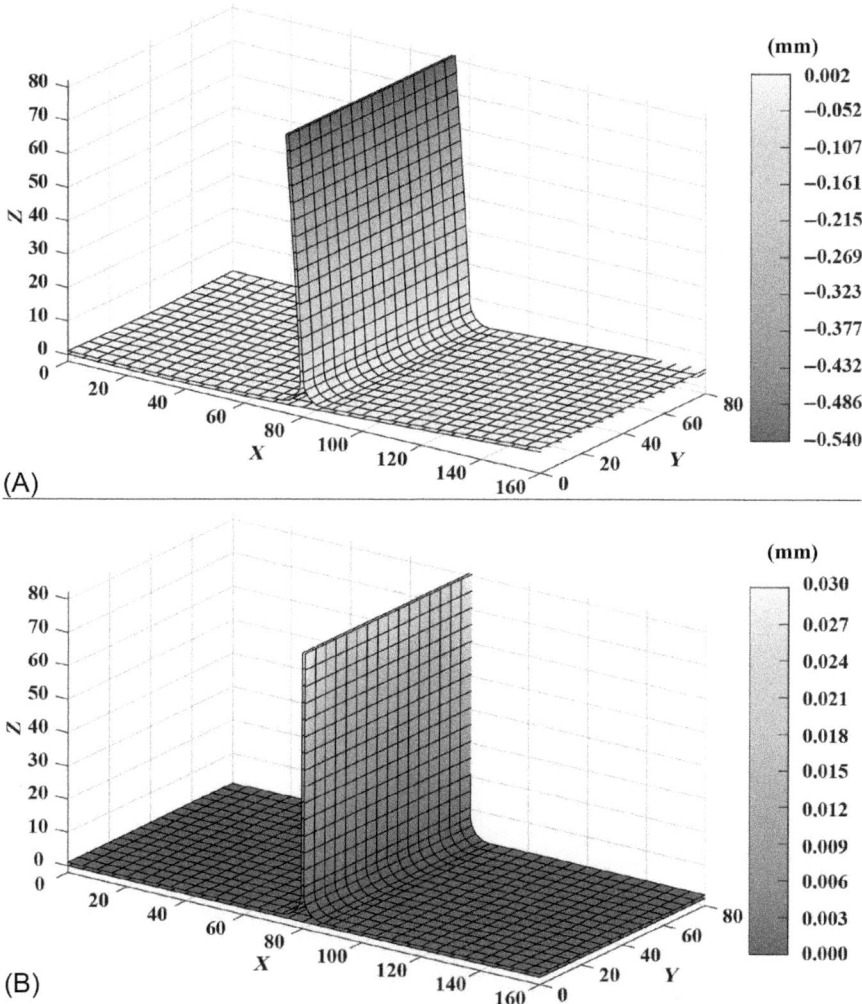

(A)

(B)

Fig. 9 Results of the case study (magnified 10 times) about assembly variation along the X-axis: (A) mean value and (B) standard deviation.

the standard deviation of the assembly are 0.548 and 0.030 mm compared with magnitudes of 1.586 and 0.080 mm before the assembly.

These maximum values and the values plotted in Fig. 10 were obtained starting from the values recorded along the three main directions (X, Y, and Z in Figs. 9, 11, and 12). So, starting from them, the global average variation, as well as the standard deviation, was evaluated by using the equations:

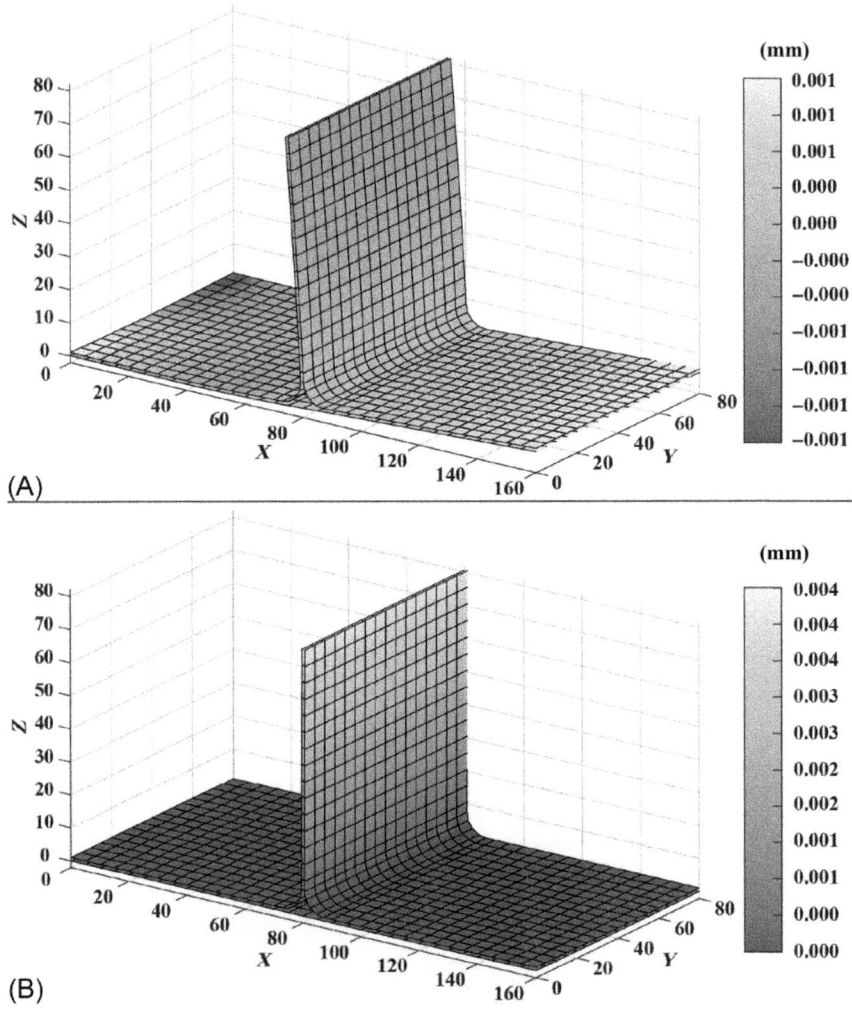

(A)

(B)

Fig. 10 Results of the case study (magnified 10 times) about resultant assembly variation: (A) mean value and (B) standard deviation.

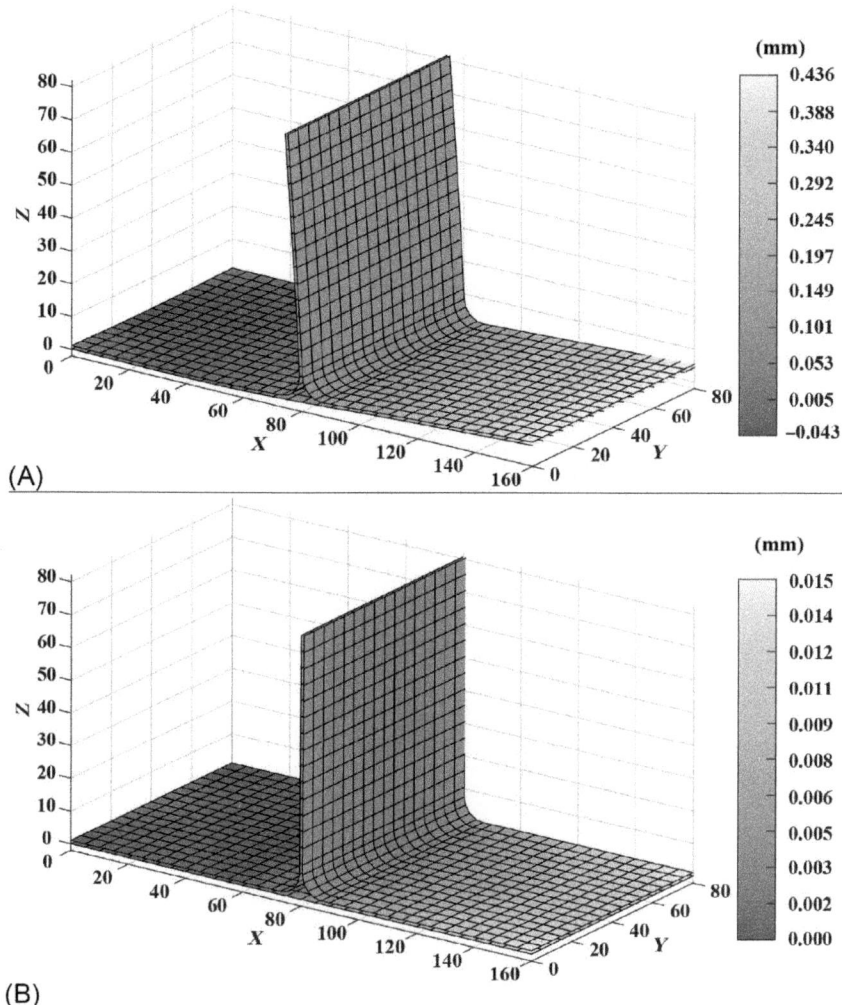

Fig. 11 Results of the case study (magnified 10 times) about assembly variation along the Y-axis: (A) mean value and (B) standard deviation.

$$\mu_a = \sqrt{\mu_{X,a}^2 + \mu_{Y,a}^2 + \mu_{Z,a}^2} \tag{14}$$

$$\sigma_a = \sqrt{\sigma_{X,a}^2 + \sigma_{Y,a}^2 + \sigma_{Z,a}^2} \tag{15}$$

where $\mu_{i,\,a}$ and $\sigma_{i,\,a}$ are the mean and standard deviation values of assembly for the ith main direction (where $i = X$, Y, and Z).

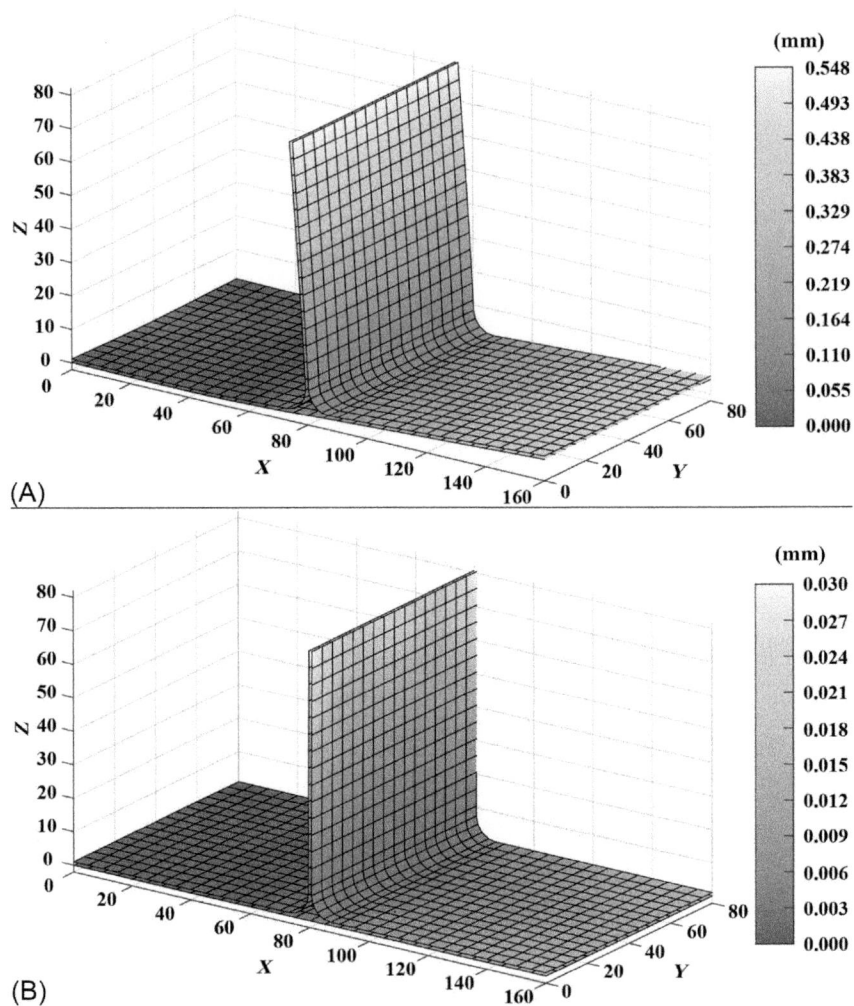

Fig. 12 Results of the case study (magnified 10 times) about assembly variation along the Z-axis: (A) mean value and (B) standard deviation.

A comparison of the results between the classical and the proposed approach is shown in Tables 2 and 3. Table 2 shows differences resulting from the classical and the proposed approach to evaluate the stiffness matrix $[K_p]$. In particular, there is a reduction of simulation time maintaining the same result. Moreover, the proposed approach involves a very small approximation that is negligible to evaluate the stiffness matrix. This error does not affect the results accuracy of the mean variation and standard deviation of

Table 2 Comparison between the classical and proposed approach to evaluate the stiffness matrix

Approach	$[K_p]$						Time (s)
Classical	3.194	−5.435	2.594	0	0	0	61
	−5.440	11.347	−5.440	0	0	0	
	2.594	−5.435	3.194	0	0	0	
	0	0	0	−3.194	5.435	−2.594	
	0	0	0	5.440	−11.347	5.440	
	0	0	0	−2.594	5.435	−3.194	
Proposed	3.195	−5.439	2.597	0	0	0	3
	−5.439	11.344	−5.439	0	0	0	
	2.597	−5.439	3.195	0	0	0	
	0	0	0	−3.195	5.439	−2.597	
	0	0	0	5.439	−11.344	5.439	
	0	0	0	−2.597	5.439	−3.195	

Percentage error (%)

Approach	$[K_p]$						Time (s)
Proposed vs classical	0.022	0.065	0.111	/	/	/	
	−0.028	−0.032	−0.028	/	/	/	
	0.111	0.065	0.022	/	/	/	
	/	/	/	0.022	0.065	0.111	
	/	/	/	−0.028	−0.032	−0.028	
	/	/	/	0.111	0.065	0.022	

Table 3 Comparison between the classical and proposed approach to evaluate the mean variation and standard deviation of assembly

Approach	Nodes	$\{\mu_a\}$	$\{\sigma_a\}$	Time (s)
Classical	860	0.509	0.021	122
	950	0.508	0.021	
	1040	0.507	0.020	
	1455	0.508	0.020	
	1545	0.510	0.021	
	1635	0.510	0.020	
Proposed	860	0.508	0.021	68
	950	0.508	0.021	
	1040	0.507	0.021	
	1455	0.508	0.020	
	1545	0.509	0.021	
	1635	0.510	0.021	
Percentage error (%)				
Proposed vs classical	860	−0.033	0.233	
	950	−0.036	−0.910	
	1040	−0.033	0.234	
	1455	−0.033	0.240	
	1545	−0.036	−0.917	
	1635	−0.033	0.238	

assembly, as shown in Table 3. Even in this case, the proposed approach allows reducing the simulation time; in particular, the complete analysis is performed in about half the time.

5 CONCLUSIONS

The aim of this work was to manage uncertainties resulting in products constituted by compliant parts that deviate from the specification. In particular, the proposed approach estimates the geometric deviations of an assembly due to the compliance of the material, to the geometric tolerances of the components, and to the fastening of the parts by involving a time smaller than that due to the classical literature approach.

The proposed approach is based on the method of influence coefficient, but uses the super-element method to rapidly solve tolerance analysis problems with a more efficient evaluation of the stiffness matrix of the bodies involved in the assembly.

The proposed approach was applied to a case study and the obtained results were compared with those due to the literature method. For this case study, there was a half-time reduction of the simulation compared to that due to the literature approach. The difference in the results in terms of geometrical deviations of the compliant assemblies was negligible.

Computational efficiency to simulate compliant assemblies is an important aspect in tolerance analysis problems, especially when geometry is too complex to involve many nodes and elements.

REFERENCES

Bourdet, P., Mathieu, L., Lartigue, C., Ballu, A., 1996. The concept of the small displacement torsor in metrology. Ser. Adv. Math. Appl. Sci. 40, 110–122.

Camelio, J., Hu, S.J., Ceglarek, D., 2003. Modeling variation propagation of multi-station assembly systems with compliant parts. J. Mech. Des. 125, 673–681. https://doi.org/10.1115/1.1631574.

Camelio, J.A., Hu, S.J., Marin, S.P., 2004. Compliant assembly variation analysis using component geometric covariance. J. Manuf. Sci. Eng. 126, 355–360. https://doi.org/10.1115/1.1644553.

Chang, M., Gossard, D.C., 1997. Modeling the assembly of compliant, non-ideal parts. Comput. Des. 29, 701–708. https://doi.org/10.1016/S0010-4485(97)00017-1.

Corrado, A., Polini, W., 2017a. Manufacturing signature in jacobian and torsor models for tolerance analysis of rigid parts. Robot Comput. Integr. Manuf. 46, 15–24. https://doi.org/10.1016/j.rcim.2016.11.004.

Corrado, A., Polini, W., 2017b. Manufacturing signature in variational and vector-loop models for tolerance analysis of rigid parts. Int. J. Adv. Manuf. Technol. 88, 2153–2161. https://doi.org/10.1007/s00170-016-8947-z.

Crisfield, M.A., 1986. Finite Elements and Solution Procedures for Structural Analysis. Pineridge Press Limited.

Davidson, J.K., Mujezinović, A., Shah, J.J., 2002. A New Mathematical Model for Geometric Tolerances as Applied to Round Faces. J. Mech. Des. 124, 609. https://doi.org/10.1115/1.1497362.

Desrochers, A., 2003. A CAD/CAM representation model applied to tolerance transfer methods. J. Mech. Des. 125, 14. https://doi.org/10.1115/1.1543974.

Desrochers, A., Rivière, A., 1997. A matrix approach to the representation of tolerance zones and clearances. Int. J. Adv. Manuf. Technol. 13, 630–636. https://doi.org/10.1007/bf01350821.

Ersoy, N., Garstka, T., Potter, K., Wisnom, M.R., Porter, D., Stringer, G., 2010. Modelling of the spring-in phenomenon in curved parts made of a thermosetting composite. Compos. Part A Appl. Sci. Manuf. 41, 410–418. https://doi.org/10.1016/j.compositesa.2009.11.008.

Falgarone, H., Thiébaut, F., Coloos, J., Mathieu, L., 2016. Variation simulation during assembly of non-rigid components. realistic assembly simulation with ANATOLEFLEX software. Procedia CIRP 43, 202–207. https://doi.org/10.1016/j.procir.2016.02.336.

Franciosa, P., Gerbino, S., Patalano, S., 2011. Advanced user-interaction with GUIs in MatLAB®. Eng. Educ. Res. Using MATLAB, InTech, https://doi.org/10.5772/21043.

Gao, J., Chase, K.W., Magleby, S.P., 1998. Generalized 3-D tolerance analysis of mechanical assemblies with small kinematic adjustments. IIE Trans 30, 367–377. https://doi.org/10.1080/07408179808966476.

Ghie, W., Laperrière, L., Desrochers, A., 2003. A unified jacobian-torsor model for analysis in computer aided tolerancing. In: Gogu, G., Coutellier, D., Chedmail, P., Ray, P.

(Eds.), Recent Adv. Integr. Des. Manuf. Mech. Eng. Springer Netherlands, Dordrecht, pp. 63–72. https://doi.org/10.1007/978-94-017-0161-7_7.

Giordano, M., Pairel, E., Samper, S., 1999. Mathematical representation of tolerance zones. In: van Houten, F., Kals, H. (Eds.), Glob. Consistency Toler. Proc. 6th CIRP Int. Semin. Comput. Toler. Univ. Twente, Enschede, Netherlands, 22–24 March, 1999. Springer Netherlands, Dordrecht, pp. 177–186. https://doi.org/10.1007/978-94-017-1705-2_18.

Gupta, S., Turner, J.U., 1993. Variational solid modeling for tolerance analysis. IEEE Comput. Graph. Appl. 13, 64–74. https://doi.org/10.1109/38.210493.

Jareteg, C., Wärmefjord, K., Söderberg, R., Lindkvist, L., Carlson, J., Cromvik, C., et al., 2014. Variation simulation for composite parts and assemblies including variation in fiber orientation and thickness. Procedia CIRP 23, 235–240. https://doi.org/10.1016/j.procir.2014.10.069.

Jayaraman, R., Srinivasan, V., 1989. Geometric tolerancing: I. Virtual boundary requirements. IBM J Res Dev 33, 90–104. https://doi.org/10.1147/rd.332.0090.

Kappel, E., Stefaniak, D., Hühne, C., 2013. Process distortions in prepreg manufacturing—an experimental study on CFRP L-profiles. Compos. Struct. 106, 615–625. https://doi.org/10.1016/j.compstruct.2013.07.020.

Laperrière, L., Lafond, P., 1999. Tolerance analysis and synthesis using virtual joints. In: van Houten, F., Kals, H. (Eds.), Glob. Consistency Toler. Springer Netherlands, Dordrecht, pp. 405–414. https://doi.org/10.1007/978-94-017-1705-2_41.

Liu, S.C., Hu, S.J., 1997. Variation simulation for deformable sheet metal assemblies using finite element methods. J. Manuf. Sci. Eng. 119, 368. https://doi.org/10.1115/1.2831115.

Lorin, S., Lindkvist, L., Söderberg, R., 2012. Simulating part and assembly variation for injection molded parts. 6th Int. Conf. Micro-Nanosyst. 17th Des. Manuf. Life Cycle Conf., ASME. Vol. 5. p. 487. https://doi.org/10.1115/DETC2012-70659.

Lorin, S., Lindkvist, L., Söderberg, R., Sandboge, R., 2013. Combining variation simulation with thermal expansion simulation for geometry assurance. J. Comput. Inf. Sci. Eng. 13, 31007. https://doi.org/10.1115/1.4024655.

Marc, M.S.C., 2013. Volume A: theory and user information.

Mortensen, A.J., 2002. An Integrated Methodology for Statistical Tolerance Analysis of Flexible Assemblies. Brigham Young University. Department of Mechanical Engineering.

Schleich, B., Anwer, N., Mathieu, L., Wartzack, S., 2014. Skin model shapes: a new paradigm shift for geometric variations modelling in mechanical engineering. Comput. Des. 50, 1–15. https://doi.org/10.1016/j.cad.2014.01.001.

Sellem, E., Rivière, A., 1998. Tolerance analysis of deformable assemblies. Proc. 1998 ASME Des. Eng. Tech. Conf., Atlanta, pp. 1–7.

Ungemach, G., Mantwill, F., 2008. Efficient consideration of contact in compliant assembly variation analysis. J. Manuf. Sci. Eng. 131, 11005. https://doi.org/10.1115/1.3046133.

Xie, K., Wells, L., Camelio, J.A., Youn, B.D., 2007. Variation propagation analysis on compliant assemblies considering contact interaction. J. Manuf. Sci. Eng. 129, 934. https://doi.org/10.1115/1.2752829.

INDEX

Note: Page numbers followed by "*f*" indicate figures, and "*t*" indicate tables.